Viv Tolley

Exam Practice
Workbook

AQA
GCSE Science B

Contents

Contents

N.B. The numbers in brackets correspond to the reference numbers on the AQA Science B specification.

1. Telescopes can be used to observe the Universe.

   (a) Name three types of terrestrial telescope, and for each one, state the type of electromagnetic radiation it can detect. (3 marks)

   (i) ....

   ....

   (ii) ....

   ....

   (iii) ....

   ....

   (b) (i) Give **one** advantage of using an optical telescope in space. (1 mark)

   ....

   (ii) Give **one** disadvantage of using an optical telescope on Earth. (1 mark)

   ....

   (c) Draw lines to match each description in **List A** with the correct type of telescope in **List B**. (4 marks)

   List A   **List B**

   | Contains a parabolic glass mirror | | Radio |

   | Orbits the Earth | | Reflecting |

   | Contains two convex (converging) lenses | | Refracting |

   | Has a large (non-glass) parabolic dish and a receiver | | Space |

   (d) Explain how a reflecting telescope works. (4 marks)

   ....

   ....

   ....

   ....

2. **(a)** Radio telescopes need to be very large compared to optical telescopes. Why is this? (1 mark)

   ....

   ....

(b) (i) Which kind of telescope uses a parabolic dish to reflect waves?   (1 mark)

....

(ii) Explain how this type of telescope works.   (4 marks)

....

....

....

....

3.   **(a)** Explain, in as much detail as you can, how a refracting telescope works.   (4 marks)

....

....

....

....

(b) Give **two** advantages and **two** disadvantages of using refracting and radio telescopes. Write your answers in the spaces in the table below.   (4 marks)

		Advantages	Disadvantages
(i)	Refracting	1. .... 2.	1. .... 2. ....
(ii)	Radio	1. .... 2.	1. .... 2. ....

(c) Why are many telescopes placed on the top of mountains and in areas with low levels of pollution?   (1 mark)

....

....

4.   **(a)** Which of the following emit light, which provides possible evidence about the origin of the Universe? Tick the correct option.   (1 mark)

Object	Tick (✓)
Planets	
Comets	
Moons	
Galaxies	

(b) The light emitted in part **(a)** displays 'red-shift'. Explain the meaning of 'red-shift'.   (2 marks)

....

....

....

....

(c) What does red-shift tell us about other galaxies in the Universe?   (3 marks)

....

....

....

....

....

....

5.   **(a)** Fill in the missing words to complete the sentences below.   (3 marks)

Light that reaches the Earth from distant stars and galaxies is shifted

.... the red end of the electromagnetic spectrum.

This is evidence that the galaxies are moving .... from

the Earth. This suggests that the Universe is ....

(b) The red-shift effect is more exaggerated in the galaxies that are furthest away. What does this suggest? (1 mark)

....

....

6. Which answer correctly describes the movement of the galaxies at the present time, **(a)** relative to the Earth, **(b)** relative to each other? Tick the correct option. (1 mark)

(a) Movement of galaxies at the present time relative to the Earth	(b) Movement of galaxies at the present time relative to each other	Tick the correct row
moving away	moving closer	
moving away	moving away	
moving closer	moving closer	
moving closer	moving away	

(Total: .... / 39 marks)

1. The Earth is a planet that has changed since its formation and is still changing.

    (a) The drawing shows the structure of the Earth.

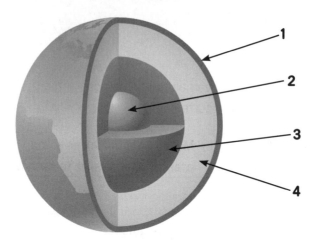

Match statements **A**, **B**, **C** and **D** with the labels **1–4** on the diagram. Enter the appropriate numbers in the boxes provided. (4 marks)

Inner core ☐

Outer core ☐

Mantle ☐

Crust ☐

    (b) What is the name of the process responsible for the rocks at the Earth's surface being continuously broken up, reformed and changed? (1 mark)

    ....

    (c) Alfred Wegener was the scientist who first suggested the idea of plate tectonics.

    (i) What evidence did Wegener use to support his theory? Tick the **three** correct options. (3 marks)

Evidence	Tick (✓)
The jigsaw fit of some continents.	
Other people believed him.	
Layers of rocks are the same on different continents.	
The Periodic Table.	
The remains of fossils.	

    (ii) Alfred Wegener proposed that the movement of the Earth's crust was responsible for separating land masses. What name did he give to this process? (1 mark)

    ....

2. The Earth consists of three layers surrounded by the atmosphere.

(a) What is the Earth's lithosphere? Tick the correct option. (1 mark)

The core and mantle ◯

The crust and upper mantle ◯

The crust and atmosphere ◯

The core, mantle and crust ◯

(b) (i) Complete the following sentence. (1 mark)

The Earth's lithosphere is cracked into a number of large pieces called ....

....

(ii) What causes the pieces of lithosphere to move? Tick the correct option. (1 mark)

Convection currents in the crust ◯

Convection currents in the mantle ◯

Convection currents in the outer core ◯

Convection currents in the inner core ◯

3. The structure of the Earth is changing all the time.

(a) Ring the correct options in the following sentence. (3 marks)

Water / Intense heat / Gas / Matter released by **combustion / currents / radioactive decay / oxygen** causes currents in the Earth's **crust / mantle / core / orbit**.

(b) Use the words provided to fill in the spaces and complete the following sentences. (3 marks)

crust	rock	slowly	radiation	down	convection	quickly

(i) Hot molten .... rises to the surface, creating new ....

(ii) The older, cooler crust then sinks .... where the ....

current starts to fall.

(iii) The land masses on these plates move ....

(c) Roughly how many major tectonic plates are there? Tick the correct option. (1 mark)

Number	Tick (✓)
3	
7	
50	
100	

4. Describe the **three** ways in which tectonic plates can move in relation to each other. (3 marks)

1. ....

....

2. ....

....

3. ....

....

5. Tectonic plates can move in different ways.

(a) Fill in the missing words to complete the sentences below. (2 marks)

(i) .... plate boundaries occur when plates collide and one is forced under the other.

(ii) .... plate boundaries occur when plates move apart and molten rock rises to the surface and forms new ocean floor.

(b) How much new ocean floor is formed each year? (1 mark)

....

6. (a) Suggest **three** natural disasters that are common at plate boundaries. (3 marks)

1. ....

2. ....

3. ....

(b) At what type of plate boundary are these natural disasters most likely to occur? (1 mark)

(c) In your own words, describe how an earthquake occurs. (3 marks)

(d) San Francisco experiences frequent earthquakes. Why do you think this is? (1 mark)

7. The Earth's atmosphere has changed a lot since its formation.

(a) When the Earth was first formed, which gas did the atmosphere mainly consist of?
Tick the correct option. (1 mark)

Oxygen ☐

Carbon dioxide ☐

Nitrogen ☐

Methane ☐

(b) Which gas does the Earth's atmosphere mainly consist of today? (1 mark)

(c) Over what time period did the change in the atmosphere take place? (1 mark)

(d) When the Earth first formed, there was intense volcanic activity. Name **three** gases released by volcanic activity. (3 marks)

1.

2.

3.

8. How do most scientists believe the oceans were created? Tick the correct option. (1 mark)

Reaction	Tick (✓)
A chemical reaction between hydrogen and oxygen.	
The water vapour in the air boiled away.	
The water vapour in the air condensed.	
Plants respiring released water vapour.	

9. **(a)** Why did the evolution of green plants produce significant changes in the Earth's atmosphere? (1 mark)

....

....

(b) Some carbon from the carbon dioxide in the air is taken out of the carbon cycle for long periods of time. What happens to this carbon? (2 marks)

....

....

....

....

10. **(a)** Ring the correct options in the following sentences. (4 marks)

(i) For about the last **100 / 200 / 300 / 400** million years the proportion of gases in the atmosphere has remained fairly constant.

(ii) There is about 80% **oxygen / nitrogen / argon / helium** and 20% **oxygen / nitrogen / argon / water** in the air.

(iii) The atmosphere also contains noble gases like **oxygen / nitrogen / carbon dioxide / helium**.

(b) **(i)** What distinguishes the noble gases from the other gases in the Earth's atmosphere? Tick the correct option. (1 mark)

They are very volatile. ◯

They are unreactive. ◯

They are not naturally occurring. ◯

They are very dense. ◯

(ii) Give **one** example of how a noble gas can be used. (1 mark)

....

11. **(a)** Fill in the missing words to complete the sentences. (4 marks)

The amount of carbon dioxide in the atmosphere is reduced by ....

in green plants.

It is also removed through a reaction with .... water, which

produces carbonates.

Insoluble carbonates are deposited as ...., which forms rocks in

the Earth's ....

(b) What are the **two** main reasons why the level of carbon dioxide in the atmosphere is
increasing? (2 marks)

(i) ....

....

(ii) ....

....

12. Recent reports suggest that the concentration of carbon dioxide in the atmosphere is increasing.
Explain how this increase in carbon dioxide may have a harmful effect on the Earth. (5 marks)

....

....

....

....

....

....

....

....

....

....

13. The composition of our atmosphere has been about the same for 200 million years. The pie charts below show the composition of the atmosphere about 4 billion years ago and the atmosphere as it is today.

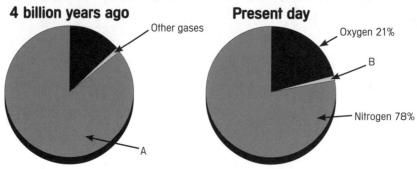

4 billion years ago
Other gases
A

Present day
Oxygen 21%
B
Nitrogen 78%

(a) (i) Name the gas found in the atmosphere 4 billion years ago (labelled A on the pie chart above). (1 mark)

_____

(ii) Describe one process that released carbon dioxide into the atmosphere 4 billion years ago. (1 mark)

_____

(b) Name one other gas, apart from oxygen or nitrogen, that is naturally present in the atmosphere today. (1 mark)

_____

(c) *In this question you will be assessed on using good English, organising information clearly and using specialist terms where appropriate.*

Describe and explain how the composition of the atmosphere has changed over the last 4 billion years. (6 marks)

_____

_____

_____

_____

_____

_____

_____

_____

_____

(Total: _____ / 69 marks)

1. Atoms are made up of subatomic particles.

   The diagram shows an atom and its subatomic particles.

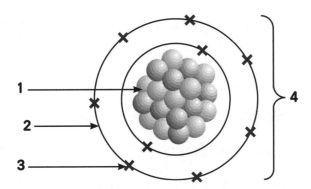

   (a) Complete the table by writing the number **1–4** from the diagram in the correct box. (4 marks)

Part of atom	Number
Electron	
Electron shell	
Nucleus	
Atom	

   (b) Fill in the missing words to complete the following sentences: (4 marks)

   (i) The smallest particle of a chemical element that can exist on its own is an ....

   (ii) Electrons are negative particles found in the .... of an atom.

   (iii) The centre of an atom is called the ....

   (iv) In an element, all the atoms are ....

2. Chemical elements can be represented by symbols.

   (a) What are the chemical symbols for the following elements? (4 marks)

   Sodium ....

   Oxygen ....

   Iron ....

   Sulfur ....

(b) What is an element? _____ (1 mark)

_____

(c) About how many elements are there in the Periodic Table? Tick the correct option. (1 mark)

    80 ◯   90 ◯   100 ◯   150 ◯

3. **(a)** Choose the correct words from the options given to complete the following sentences. (2 marks)

          electrons   **protons**   **neutrons**

    (i) An atom contains an equal number of _____ and _____.

    (ii) The nucleus contains _____ and _____, and is

        surrounded by _____.

(b) Draw lines between the boxes to match each atomic particle to its charge. (3 marks)

Neutron		−1
Proton		0
Electron		+1

(c) Why do atoms have no charge? Tick the correct option. (1 mark)

They contain the same number of electrons and protons ◯

They contain the same number of electrons and neutrons ◯

They contain the same number of protons and neutrons ◯

They contain the same number of protons, neutrons and electrons ◯

4. Draw lines between the boxes to match each key word in **List A** to its definition in **List B**. (4 marks)

List A   **List B**

List A	List B
Molecule	The simplest particle that can exist on its own
Element	All of the atoms are the same in these substances
Atom	A substance formed when two or more different types of atom are chemically joined together
Compound	A particle that consists of two or more atoms chemically joined

5. The diagram shows the electron configuration of an atom.

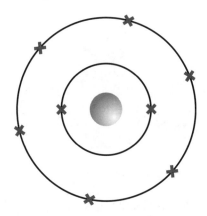

(a) What is the proton (atomic) number of this atom? (1 mark)

_____

(b) What group of the Periodic Table does this atom belong to? (1 mark)

_____

(c) What period of the Periodic Table does this atom belong to? (1 mark)

_____

(d) What is the electron configuration of this atom? (1 mark)

_____

(e) What would be the charge of an ion made from this atom? Tick the correct option. (1 mark)

−2 ◯   −1 ◯

+1 ◯   +2 ◯

6. Explain how different atoms join together to make molecules and compounds. (3 marks)

_____

_____

_____

_____

_____

_____

7. This diagram shows an ammonia molecule.

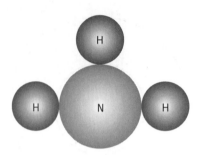

Atoms can bond with each other to form molecules.

(a) How many atoms are there in this molecule? (1 mark)

....

(b) How many elements are there in this molecule? (1 mark)

....

(c) What is the chemical formula of the compound shown? (1 mark)

....

8. Complete the table to show the number of molecules and the number of different elements represented by the chemical formulae. Write your answers in the empty spaces in the table. (4 marks)

Chemical formula	Number of molecules	Number of different elements
NaOH		
3ZnO		
$2H_2O$		
$4H_2SO_4$		

9. The full Periodic Table lists all the known elements.

(a) Fill in the missing words to complete the following sentence. (2 marks)

Whatever version of the Periodic Table is used, for each element the ....

number is always the top number and the .... number is always at the bottom.

(b) Circle the correct options in the following sentences. (2 marks)

The mass number tells you the number of **protons / electrons** and neutrons in the nucleus.

The atomic number is the number of **protons / electrons / neutrons** in the nucleus.

(c) Use the copy of the Periodic Table at the back of this workbook to answer this section.

(i) How many protons does an oxygen atom have? Tick the correct option. (1 mark)

1 ◯   16 ◯

8 ◯   0 ◯

(ii) Which element has six protons? Tick the correct option. (1 mark)

B ◯

C ◯

N ◯

O ◯

(iii) What is the mass number of sodium? (1 mark)

(iv) How many neutrons does a sodium atom have? (1 mark)

(v) Which of the statements below about the mass of particles in an atom is true?
Tick the correct option. (1 mark)

Statement	Tick (✓)
A proton has the same relative mass as a neutron	
A proton has the same relative mass as an electron	
A neutron has the same relative mass as an electron	
A proton, a neutron and an electron all have the same relative mass	

10. Atoms contain different numbers of sub-atomic particles.

(a) Use the copy of the Periodic Table at the back of this workbook to help complete the table below. (4 marks)

	14 N 7	197 Au 79	235 U 92	(i) .... Ca (ii) ....
Number of protons	(i)	(ii)	(iii)	20
Number of neutrons	(i)	(ii)	(iii)	20

(b) Some elements have different isotopes. Explain the meaning of the word **isotope**. (2 marks)

....

....

....

11. Use the Periodic Table at the back of this workbook to help you answer these questions.

Hydrogen and helium are elements found in the Periodic Table. The following are symbol representations of two isotopes of hydrogen.

A 1
 H
1

B 2
 H
1

(a) How do we know that they are isotopes of hydrogen? (1 mark)

....

(b) How many electrons would isotope **A** contain? (1 mark)

....

(c) How many neutrons would isotope **B** contain? (1 mark)

....

(d) What is the relative atomic mass of helium? (1 mark)

....

(e) Complete the following sentence. (1 mark)

The relative atomic mass is an average value for all the .... of the element.

12. In the nucleus of a potassium atom there are 19 protons and 20 neutrons.

(a) What is the mass number of potassium? (1 mark)

....

(b) What is the atomic number of potassium? (1 mark)

....

(c) How many electrons does an atom of potassium contain? (1 mark)

....

(d) Why is the atom of potassium neutral in terms of charge? (1 mark)

....

13. **(a)** What do the fractions of crude oil contain? Tick the **three** correct options.　(3 marks)

Options	Tick (✓)
Mixture of compounds	
Separate atoms	
Hydrocarbons	
Alkanes	
Each fraction is a pure compound	

(b) Ring the correct options in the following sentences:　(2 marks)

Fractions with low boiling points come out at **the top / middle / bottom** of the fractionating column.

Fractions with high boiling points come out at the **top / middle / bottom** of the fractionating column.

(c) Petrol has a relatively low boiling point. Is petrol a long or short-chain hydrocarbon?　(1 mark)

14. Metals have many uses.

(a) What process is used to extract aluminium from its ore?　(1 mark)

(b) Titanium is a very lightweight and strong metal.
Which of the following are uses for titanium? Tick the correct option(s).　(2 marks)

Aircraft ◯ Drinks cans ◯

Replacement joints ◯ Sunscreen ◯

(c) Aluminium is a good conductor of electricity, and is flexible and light. Which of the following is **not** used for aluminium? Tick the correct option.　(1 mark)

Sandwich wrapping ◯ Overhead power lines ◯

Water pipes ◯ Drinks cans ◯

(d) Circle the correct options in the following sentences:　(3 marks)

Unreactive metals that are found in the Earth's crust are said to be native. An example of a native

metal is **gold / calcium**.

Most metals occur as **nuggets / compounds** in the Earth's crust.

If the amount of the metal is sufficient for it to be economic to extract it from these rocks, they are

also known as **minerals / ores**.

(e) Name three metals that can be found native in the Earth's crust. (3 marks)

15. Iron metal can be removed from iron ore in a blast furnace.

(a) What type of chemical reaction happens in a blast furnace to produce pure iron from iron oxide? (1 mark)

(b) Choose the correct word from the options given to complete the sentence below: (1 mark)

hard	soft	brittle	shiny

Pure iron is used for construction because it is ....

(c) What substances are added to iron to make steel? (2 marks)

16. Crude oil is a mixture of different molecules.

(a) What is a mixture? Tick the correct option. (1 mark)

Options	Tick (✓)
More than one atom chemically joined	
More than one type of atom not chemically joined	
More than one type of substance not chemically joined	
More than one type of substance chemically joined	

(b) What is crude oil a mixture of? Tick the correct option. (1 mark)

Options	Tick (✓)
A mixture of hydrocarbon molecules	
A mixture of carbon molecules	
A mixture of hydrogen and carbon	
More than one type of substance chemically joined	

(c) Is the following sentence **true** or **false**?

The longer the hydrocarbon chain in a molecule, the higher its boiling point. (1 mark)

_____

(d) Crude oil can be separated into fractions by fractional distillation. Explain how fractional distillation happens. (3 marks)

_____

_____

_____

_____

_____

_____

17. **(a)** What does M_r stand for? Tick the correct option. (1 mark)

The element manganese ⬭   The relative atomic mass ⬭

The atomic number ⬭   The relative formula mass ⬭

(b) What is the M_r of calcium carbonate, $CaCo_3$? (2 marks)

_____

(c) When calcium carbonate is heated it undergoes thermal decomposition to produce calcium oxide and carbon dioxide.

Write a word equation for this reaction. (1 mark)

_____

(d) The formula for calcium carbonate is $CaCO_3$. Which of the following is the correct symbol equation for the thermal decomposition of calcium carbonate? Tick the correct option. (1 mark)

$CaCO_3 \longrightarrow 2CaO + CO_2$ ⬭

$CaO \longrightarrow CaO + CO_2$ ⬭

$CaCO_3 \longrightarrow CaO + CO_2$ ⬭

$CaCO_3 \longrightarrow CaO + CO_2$ ⬭

(Total: _____ / 94 marks)

Using Materials from Our Planet to Make Products

U1

1. **(a)** Name the reactant(s) and the product(s) in the following reaction. (2 marks)

magnesium + oxygen ⟶ magnesium oxide

Reactant(s): ....

Product(s): ....

(b) Stephen and Joe are investigating the reactions of magnesium, iron and copper with sulfuric acid.

Dilute Sulfuric Acid

(i) What does the hazard symbol on the bottle mean? (1 mark)

....

(ii) Describe one safety precaution Stephen and Joe should take before they start their investigation. (1 mark)

....

(c) Stephen and Joe place a small amount of each metal in a test tube. Next they add sulfuric acid to each test tube. Then they observe each tube and record the number of bubbles produced.

1   2   3

(i) Name the independent variable in this investigation. (1 mark)

....

(ii) Describe one variable they should control. (1 mark)

....

(iii) How would a change in this variable affect the results? (1 mark)

....

(d) The table shows the results of the investigation.

(i) Complete the table to show which metal was in each tube. (3 marks)

Test tube	Number of bubbles collected in 1 minute	Name of metal
1	23	
2	8	
3	0	

(ii) Use the results to complete the graph. The results for test tube 2 have been plotted for you.

(2 marks)

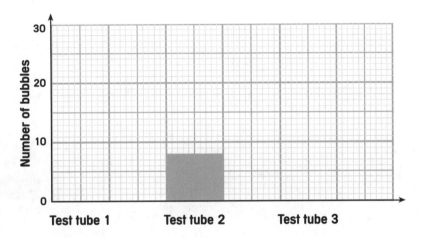

(e) Pure aluminium is separated from aluminium oxide using electricity.

(i) Explain why the ore is melted. (2 marks)

....

....

(ii) Explain how the electrical energy separates the aluminium metal from oxygen. (1 mark)

....

....

(iii) Obtaining aluminium metal is an expensive process. Use information from this question to explain why. (2 marks)

....

....

2. Some metals are more reactive than other metals. More reactive metals can displace less reactive metals from solutions. Cheryll and Russell were asked to investigate the reactivity of three metals – iron, copper and calcium. They were given small pieces of each metal and three solutions that contained a metal salt.

Cheryll and Russell set up their apparatus as shown below.

They added the same mass of:

- iron metal to all the test tubes labelled 1
- copper to all the test tubes labelled 2
- calcium to all the test tubes labelled 3

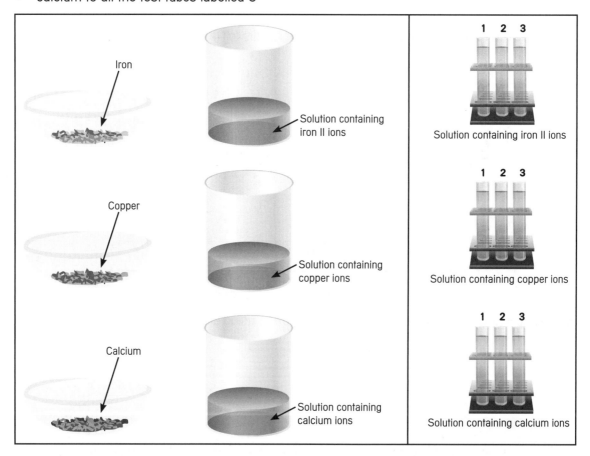

They recorded their observations in a table.

Solution containing metal ions	Iron metal added	Copper metal added	Calcium metal added
Iron	No reaction	No reaction	Metal deposited at the bottom of the test tube
Copper	Metal deposited at the bottom of the tube	No reaction	Metal deposited at the bottom of the test tube
Calcium	No reaction	No reaction	No reaction

(a) Describe what these results show. (3 marks)

(b) Explain these results. (3 marks)

3. **(a)** Metals have many uses.

Describe **one** use for each of these metals. (2 marks)

Aluminium: ....

Gold: ....

(b) Metals are good conductors of heat. Explain why. (2 marks)

(Total: .... / 27 marks)

4. Fill in the missing numbers to balance the equations below.

(a) CH........... + $2O_2$ ⟶ CO........... + $2H_2O$ (2 marks)

(b) N_2 + ...H_2 ⟶ ....NH_3 (2 marks)

5. Fill in the missing chemical symbols to balance the equations below.

(a) $CuCO_3$ ⟶ ....O + CO_2 (1 mark)

(b) 2H............ + CuO ⟶ $CuCl_2$ + H_2O (1 mark)

6. Sulfuric acid (H_2SO_4) can react with calcium hydroxide ($Ca(OH)_2$) to form calcium sulfate ($CaSO_4$) and water (H_2O). Fill in the missing chemical symbols to balance the equation for the following chemical reaction.   (4 marks)

............$(OH)_2$ + ...$_2SO_4$ ⟶ Ca............O_4 + $2H_2$............

7. Oxygen can exist as a diatomic molecule, i.e. two oxygen atoms joined together.

Use the information above to write a balanced equation for the following reaction: Copper (Cu) and oxygen (O) react to form copper oxide (CuO).   (2 marks)

....

....

....

(Total: .... / 12 marks)

1. Living organisms often have to compete with each other to survive.

(a) Which of the following statements is the correct meaning of the word **population**? Tick the correct option. (1 mark)

The total number of organisms living in a particular habitat. ◯

The total number of individuals of the same species living in a particular habitat. ◯

The total number of animals living in a particular habitat. ◯

The total number of plants living in a particular habitat. ◯

(b) Which of the following statements describes a community? Tick the correct box. (1 mark)

A group of animals and plants interacting with one another. ◯

A population of animals. ◯

Animals adapted to their surroundings. ◯

A food chain. ◯

(c) Which of the following do animals compete for? Tick the correct choices. (1 mark)

Light ◯   Food ◯   Water ◯   Space ◯

(d) Name one thing that plants compete for but animals do not. (1 mark)

....

(e) Use the words from the box to complete the sentences. (3 marks)

| survive | die | better adapted | smaller | larger |

When organisms compete, those which are ....

to their environment are more likely to .... and usually exist in

.... numbers.

(f) What would you expect to happen to the population size of an organism that was less well adapted to its environment? Explain your answer. (2 marks)

....

....

....

2. Animals and plants are found in many different environments. They have adaptations.

   (a) Which of the following statements best describes what the word **adaptation** means?
   Tick the correct option.   (1 mark)

   A significant change in the size of a population.   ⬭

   A feature that develops to make an organism better suited to its environment. ⬭

   The process by which the different organisms in a habitat learn to share resources. ⬭

   A characteristic that is acquired by an individual, e.g. a scar.   ⬭

   (b) What is the advantage of an adaptation?   (1 mark)

   ....

   (c) Polar bears live in the Arctic where it is very cold. There is snow on the land all year round and ice forms over the sea.

   Suggest **two** adaptations a polar bear might have. Explain how these adaptations help the polar bear to survive in these cold conditions.   (4 marks)

   ....

   ....

   ....

   ....

   ....

   ....

   ....

   ....

   ....

   ....

(d) Cacti have spines instead of leaves. How do these spines help them to survive?   (3 marks)

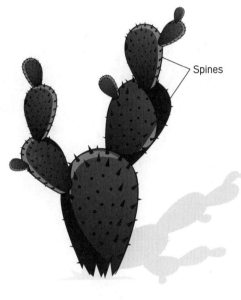

Spines

....
....
....
....
....
....
....

3.   **(a)** When do we think life forms might first have existed on Earth? Tick the correct option.   (1 mark)

300 years ago ☐   3 million years ago ☐

3 billion years ago ☐

(b) What provides evidence to support the theory of evolution? Tick the correct option.   (1 mark)

Animals ☐   Fossils ☐

Plants ☐   Viruses ☐

(c) Fill in the spaces in the following sentences by choosing words from the box below.   (3 marks)

population	species	habitat	adapted

Evolution is the change in a .... over many generations. It may result in

formation of a new ...., the members of which are better

.... to their environment.

(d) Viv and Chris observed a population of butterflies. They noticed that most butterflies had white-coloured wings. One butterfly had brown-coloured wings. Wing colour is controlled by genes.

Suggest what might have caused this different wing colour.   (3 marks)

....

....

....

....

....

4.   Scientists have observed that, over many years, there has been a change in the population of a species of fly, where the number of dark-coloured flies has become greater than the number of light-coloured flies.

The flies are found on the trunks of trees. Over many years, the trunks of the trees have become darker in colour.

Many years ago   Now

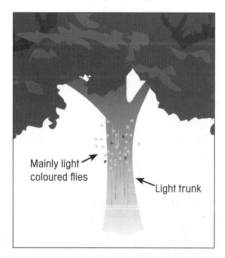

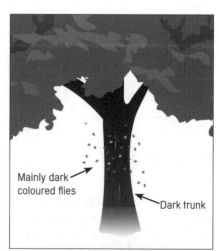

Mainly light coloured flies

Light trunk

Mainly dark coloured flies

Dark trunk

(a) *In this question you will be assessed on using good English, organising information clearly and using specialist terms where appropriate.*

Explain how the change has taken place using your knowledge about evolution. (6 marks)

_____

_____

_____

_____

_____

_____

_____

_____

_____

_____

(b) Give **three** factors that could contribute to the extinction of a species. (3 marks)

_____

_____

_____

_____

_____

(Total: .... /35 marks)

1. **(a)** The diagram below represents a food chain.

    Rosebush ⟶ Aphid ⟶ Ladybird ⟶ Blackbird

    (Ring) the correct options in the following sentences.   (3 marks)

    (i) The producer in the food chain is the **aphid / blackbird / ladybird / rosebush**.

    (ii) The herbivore in the food chain is the **aphid / blackbird / ladybird/ rosebush**.

    (iii) The top carnivore in the food chain is the **aphid / blackbird / ladybird / rosebush**.

    (b) Where does the initial source of energy for all food chains come from?   (1 mark)

    ....

    (c) The diagram shows part of a food chain.

    Blackbird ⟶ Cat

    Some energy taken in by the blackbird is not transferred to the cat. Give **two** examples of what
    happens to the energy that is not transferred to the cat.   (2 marks)

    1. ....

    2. ....

    (d) Farmers who rear animals for food try to increase the percentage of energy transferred to the next
    stage in the food chain by controlling conditions around them. The diagram shows part of a food chain.

    Cow food pellets ⟶ Cows

    Suggest **two** methods that could be used to improve the percentage of energy transferred
    through this part of the food chain. Explain how each method increases the amount of energy
    transferred.   (4 marks)

    ....

    ....

    ....

    ....

    ....

2. Briefly explain what biomass is.   (1 mark)

    ....

    ....

    (Total: .... / 11 marks)

1. Materials such as carbon, nitrogen, hydrogen and oxygen are 'recycled' in the environment through cycles.

(a) Fill in the spaces in the following sentences by choosing words from the box below.   (4 marks)

growth	die	returned	remove	reproduce	add	live

Organisms .... material from the environment for ....

and other purposes. These materials are .... to the environment when the

organisms ....

(b) Give **three** factors that increase the rate at which microorganisms 'recycle' materials.   (3 marks)

1. ....   **2.** ....   **3.** ....

(c) The diagram below shows the carbon cycle.

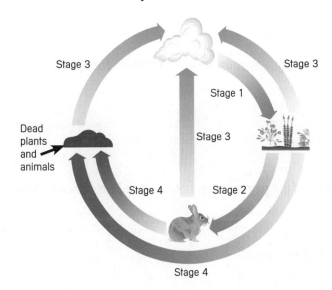

(i) Using the diagram, circle the correct options in the following sentences.

Stage 1 shows when **respiration / decomposition / photosynthesis / combustion** happens. (1 mark)

Stage 3 shows when **respiration / decomposition / photosynthesis / combustion** happens. (1 mark)

Stage 4 shows when **respiration / decomposition / photosynthesis / combustion** happens. (1 mark)

(ii) At stage 2, carbon is converted into certain substances in the animal. What are these
substances? Tick the correct option.   (1 mark)

Vitamins and minerals   ◯

Carbohydrates and proteins   ◯

Fats and fibre   ◯

Carbohydrates, fats and proteins   ◯

2. Plants recycle materials during growth.

(a) Fill in the spaces in the following sentences by choosing words from the box below. (3 marks)

molecules	soil	leaves	ions	flowers	roots	stem

Plants need mineral ...., which they absorb from the ....

through their ....

(b) Plants use nitrates to make large molecules. Name **two** of these molecules. (2 marks)

....

....

(c) Complete this sentence. (1 mark)

Plants use magnesium to make ....

(d) Plants can carry out a process called photosynthesis. Write a word equation to represent this process. (2 marks)

....

....

(e) (i) What is the name of the pigment that absorbs the Sun's energy during photosynthesis? (1 mark)

....

(ii) Where is this pigment found? (1 mark)

....

3. Which of the following factors does **not** limit the rate of photosynthesis? Tick the correct option. (1 mark)

Amount of oxygen ◻

Amount of carbon dioxide ◻

Temperature ◻

Amount of light ◻

Amount of chlorophyll ◻

4. **(a)** A plant is receiving plenty of light but its rate of photosynthesis stops increasing.

What other factors might be responsible? Tick the correct options. (2 marks)

Amount of carbon dioxide ☐

The temperature ☐

Amount of chlorophyll ☐

Amount of oxygen ☐

(b) Explain why too little light can reduce the growth of a plant. (2 marks)

....

....

....

....

(c) Market gardeners grow tomatoes to sell. The faster the tomato plants grow, the more money the market gardener can make. The plants are grown in buildings.

Explain how growing the tomato plants in buildings, where conditions can be controlled, will increase the growth of the plants. (4 marks)

....

....

....

....

....

....

....

....

(Total: .... / 30 marks)

1. Name the **four** main parts of the nervous system. (4 marks)

   1. ....

   2. ....

   3. ....

   4. ....

2. Identify the parts of the nervous system shown on the diagram. (3 marks)

   (a) ....

   (b) ....

   (c) ....

3. **(a)** Explain what a neurone is and what its function is. (1 mark)

   ....

   ....

   (b) Describe **two** ways in which the structure of the neurone is adapted to carry out this function.

   (i) .... (2 marks)

   ....

   (ii) .... (2 marks)

   ....

4. Which of the following structures contains sensory receptors? Tick the correct option. (1 mark)

   The liver ◯   The skin ◯

   The kidneys ◯   The stomach ◯

5. The sentences below describe what happens when an electrical impulse travels along two neurones. The first neurone is called neurone A and the second one is called neurone B. Complete the sentences using the words in the box. You can use each word more than once. (5 marks)

synapse	hormone	electrical impulse	receptor
chemical transmitter		neurone	brain

An electrical impulse travels along neurone A until it reaches a gap called the

_____. Neurone A releases a _____. This travels across the

gap and activates a _____ on neurone B. In neurone B, an _____

is generated. The _____ is immediately destroyed.

6. Impulses transfer information between receptors and effectors.

(a) Fill in the missing words to complete the following sentences. (3 marks)

The _____ neurones receive impulses from the receptors and send them to the central nervous system.

The _____ neurones send impulses from the central nervous system to the effectors.

These two neurones are connected in the central nervous system by a

_____ neurone.

(b) 'An electrical signal only travels in one direction down a neurone'.
Is this statement **true** or **false**? (1 mark)

_____

(c) What is the name of the gap between two nerve cells? (1 mark)

_____

(d) What happens when an electrical, or nervous, impulse reaches the gap between two neurones?
Tick the correct option. (1 mark)

Reaction	Tick (✓)
The impulse stops	
The impulse jumps the gap	
A chemical transmitter is released	
A hormone is released	

7. What is another name given to nerve cells? Tick the correct option. (1 mark)

Capillaries ☐   Effectors ☐

Neurones ☐   Receptors ☐

8. (a) Which of the following shows the correct sequence of events in the passage of a nerve impulse? Tick the correct option. (1 mark)

Sequence	Tick (✓)
Effector → Receptor → Sensory neurone → CNS → Motor neurone	
Receptor → Sensory neurone → CNS → Motor neurone → Effector	
Receptor → Motor neurone → CNS → Sensory neurone → Effector	
Effector → Sensory neurone → CNS → Motor neurone → Receptor	

(b) Which part of the nervous system acts as the coordinator in the passage of a nerve impulse? Tick the correct option. (1 mark)

Option	Tick (✓)
Effector	
Receptor	
Brain	
Synapse	

(c) Our bodies respond to danger very quickly to prevent injury. The sentences below describe one way in which the body responds to danger.

Fill in the missing words to complete the following sentences. (4 marks)

In a .... action the electrical impulse does not enter the conscious areas of

your brain. The time between the stimulus and the .... is as short as possible.

The only three neurones involved in this type of action are the sensory neurones, the

.... neurones and the .... neurones.

9. **(a)** Sometimes a conscious action is too slow to prevent harm to the body. A reflex action speeds up the response time by missing out the brain completely.

Which of the following is an example of a reflex action? Tick the correct option. (1 mark)

Looking both ways to cross the road ⬭

Removing your hand from a hot plate ⬭

Laughing at a joke ⬭

Whistling to get someone's attention ⬭

(b) Describe how a reflex action happens in the body. (3 marks)

....

....

....

....

....

....

(c) Explain the importance of a reflex action to the body. (3 marks)

....

....

....

....

....

....

....

10. Our nervous system detects changes around us and helps us to react to the changes.

(a) Name **three** receptors and the stimuli they detect.

Receptor: .... **Stimulus:** .... (2 marks)

Receptor: .... **Stimulus:** .... (2 marks)

Receptor: .... **Stimulus:** .... (2 marks)

(b) What is the function of the brain as part of the nervous system? (1 mark)

....

(c) Jon and Fred investigated their reaction times using a reaction time ruler.

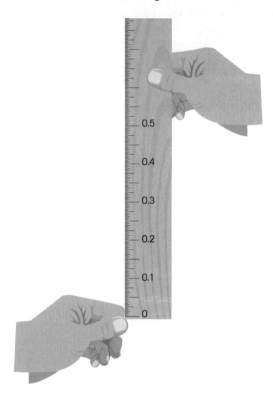

Fred held the ruler vertically. Jon put his finger and thumb level with the 0 line on the bottom of the ruler. Fred dropped the ruler and Jon had to catch it as quickly as he could. They repeated the test five times and recorded their results in a table.

Test	Time taken to catch the ruler (s)
1	0.12
2	0.36
3	0.37
4	0.35
5	0.32

(i) Which result is anomalous? (1 mark)

_____

(ii) Calculate the mean of these results. Show your working. (2 marks)

_____

_____

Mean reaction time .... s.

(d) Most people can concentrate better when they are in a quiet room.

Describe how the students could investigate the effect of noise on their reaction times.   (4 marks)

(e) Using a mobile phone when driving is against the law and dangerous. Explain why.   (2 marks)

11. **(a)** Explain the meaning of homeostasis.   (1 mark)

(b) The amount of water in the body must be controlled. Suggest **three** ways in which water is removed from the body.   (3 marks)

1.

2.

3.

(c) How does the body take in ions? Tick the correct option.   (1 mark)

By sweating ☐

By breathing ☐

By eating and drinking ☐

By excretion ☐

12. **(a)** Where are hormones produced? (1 mark)

_____

(b) How are hormones transported around the body? Tick the correct option. (1 mark)

By the nervous system ◯   By the skin ◯

In the air we breathe in ◯   In the bloodstream ◯

(c) Where is the pituitary gland? Tick the correct option. (1 mark)

Brain ◯   Liver ◯

Kidneys ◯   Heart ◯

(d) What hormone is produced by the ovaries? (1 mark)

_____

(e) What hormone is produced by the pancreas? (1 mark)

_____

13. **(a)** Which sentence best describes diabetes? Tick the correct option. (1 mark)

Pancreas does not produce insulin ◯

Liver does not produce glycogen ◯

Kidneys do not remove glucose from the blood ◯

Liver does not produce insulin ◯

(b) Which **two** hormones work together to maintain a constant glucose concentration in the body? (2 marks)

_____

_____

14. **(a)** Where are the receptors located that provide information about blood temperature? Tick the correct option. (1 mark)

Skin ◯

Heart ◯

Kidneys ◯

Lungs ◯

(b) Circle the correct option in the following sentence. (1 mark)

The normal body temperature is **20°C / 100°C / 37°C / 75°C**.

(c) What changes occur if the body becomes too cold? Tick the **three** correct options. (3 marks)

Blood vessels in the skin dilate ⬭   Skin becomes flushed ⬭

Sweat glands stop producing sweat ⬭   Blood vessels in the skin constrict ⬭

Shivering occurs ⬭   Heat loss increases ⬭

(d) Give **two** changes that occur if the body becomes too hot. (2 marks)

1. ....

2. ....

15. The blood glucose concentration in our bodies is kept constant.

(a) Name the organ that monitors the blood glucose concentration. (1 mark)

....

(b) *In this question you will be assessed on using good English, organising information clearly and using specialist terms where appropriate.*

Describe how the two hormones insulin and glucagon keep the blood glucose concentration constant. (6 marks)

....

....

....

....

....

....

....

....

....

....

(Total: .... / 81 marks)

1. Our bodies contain chemicals that are acids and alkalis.

(a) Fill in the missing words in the paragraph below. (2 marks)

When an .... and an .... are added together in the

correct amounts they 'cancel out' each other. This type of reaction is called neutralisation.

(b) (i) What is a salt? (2 marks)

....

....

(ii) A salt can be made in several ways. A salt usually has a metal and a non-metal part. Describe

three ways in which a salt can be made. (3 marks)

1. ....

2. ....

3. ....

2. Which of the following metals reacts most violently with acid? Tick the correct option. (1 mark)

Metal	Tick (✓)
Silver	
Magnesium	
Zinc	
Lead	

3. Fill in the missing words to complete the following sentences. (2 marks)

(a) .... are the oxides and hydroxides of metals.

(b) Soluble bases are called ....

4. (a) Complete the following word equation: (1 mark)

acid + base ⟶ .... + water

(b) What salt is produced when sodium hydroxide is reacted with nitric acid? (1 mark)

....

(Total: .... / 12 marks)

1. **(a)** Label the diagram of an animal cell using the words from the box. Write the appropriate words next to the letters **A–D** below. (4 marks)

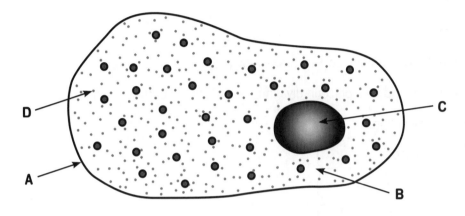

| cell membrane | nucleus | mitochondrion | cytoplasm |
| cell sap | chloroplast | cell wall | |

A ....

B ....

C ....

D ....

(b) The following definitions describe parts of an animal cell. Name the parts of the cell described in **(i–iv)**, below. (4 marks)

(i) It contains genetic information.

....

(ii) It controls the movement of substances in and out of the cell.

....

(iii) Chemical reactions take place in this part of the cell.

....

(iv) These structures are found in pairs in all body cells.

....

2. **(a)** What is variation? (1 mark)

....

(b) Name **two** causes of variation and give a named example for each cause.

    (i) Cause: .... (2 marks)

       Example: ....

    (ii) Cause: .... (2 marks)

       Example: ....

3. Mitosis is the division of body cells to make new cells.

    (a) When does mitosis take place in dividing cells? Tick the correct options.   (1 mark)

       Asexual reproduction   ◯

       Gamete production   ◯

       Repair of damaged cells   ◯

       Growth of new cells   ◯

    (b) Fill in the spaces in the following sentences by choosing words from the box below.   (3 marks)

chromosome	gene	parent	chloroplast	genetic	asexual

    A copy of each .... is made before a cell divides. The new cell has the same

    .... information as the .... cell.

    (c) Circle the correct option in the following sentence.   (1 mark)

    When one cell has undergone mitosis, **1 / 2 / 4 / 8** 'daughter' cells will be made.

4. During fertilisation the male and female sex cells join. Each sex cells contains 23 chromosomes.

    (a) How many pairs of chromosomes does the new human body cell contain?   (1 mark)

    ....

    (b) What happens to the new body cell after the sex cells have joined?   (2 marks)

    ....

    ....

    ....

    (c) Circle the correct option in the following sentence.   (1 mark)

    Human body cells contain a total of **23 / 46 / 22 / 28** chromosomes.

(d) What are sex cells known as? Tick the correct option. (1 mark)

Genes ◯ Alleles ◯

Gametes ◯ Chromosomes ◯

(e) What do sex cells contain? Tick the correct option. (1 mark)

Half the number of chromosomes as a normal body cell ◯

The same number of chromosomes as a normal body cell ◯

Twice the number of chromosomes as a normal body cell ◯

Half the number of chromosomes of a sperm cell ◯

(f) What is produced from the fusion of two sex cells? (1 mark)

....

5. **(a)** Choose the correct words from the options given to complete the following sentence. (3 marks)

oviduct **ovaries** **chromosomes** **eggs and sperm**
genes **gametes** **sperm duct**

Meiosis takes place in the .... and testes, and produces

.... containing 23 ....

(b) What is the name of the cells produced after meiosis? (1 mark)

....

6. **(a)** Choose the correct chromosomes from the options given to complete the following sentence. (1 mark)

| XY and YY | XX and XY | XX and YY | XF and XM |

The sex chromosomes in humans are represented by ....

(b) Which of the following shows female sex chromosomes? (1 mark)

A   B

Answer: ....

(c) What determines the gender of an individual? (1 mark)

....

....

7. **(a)** How many alleles does the gene controlling eye-colour have?
Tick the correct option. (1 mark)

One ☐ Four ☐

Three ☐ Two ☐

(b) Fill in the missing words to complete the following sentence. (2 marks)

Where there are two different alleles for a gene, one is known as the .... allele

and the other is known as the .... allele.

(c) Using the correct genetic terms, describe the following genotypes in terms
of their alleles. (3 marks)

(i) BB ....

(ii) Bb ....

(iii) bb ....

8. Match definitions **A–D** with the keywords **1–4** listed below. Write the appropriate numbers in the boxes
provided. (4 marks)

1 Dominant

2 Phenotype

3 Heterozygous

4 Homozygous

A What the organism looks like ☐

B The stronger allele ☐

C Both alleles are the same ☐

D Different alleles ☐

9. **(a)** Fill in the missing words to complete the following sentences. (2 marks)

(i) A .... allele will control the characteristics of the gene if it's present on
only one chromosome.

(ii) A .... allele will only control the characteristics of the gene if it is present
on both chromosomes.

(b) What does a genetic cross diagram show about the individuals? (4 marks)

10. **(a)** John has blue eyes. Both his parents have brown eyes. His mother's genotype is Bb. What must John's father's genotype be? Tick the correct option. (1 mark)

Bb ◯

BB ◯

bb ◯

(b) Circle the correct option in the following sentence. (1 mark)

If both parents have blue eyes there is a **50% / 100% / 0% / 24%** chance that they will have a child with brown eyes.

(c) (i) Complete this genetic diagram to show the possible genotypes of the offspring. (3 marks)

Brown eyes   x   Blue eyes

Parents   BB   x   bb

Gametes   ◯ ◯   ◯ ◯

Offspring   ◯ ◯   ◯ ◯

(ii) Complete the following sentence. (1 mark)

There is a .... % chance that the offspring will have brown eyes.

11. **(a)** What is an inherited disease? Tick the correct option. (1 mark)

Disease	Tick (✓)
A disease caused by microbes	
A disease passed from person to person	
A disease passed on from parent to child by genes	
A self-inflicted disease	

(b) Fill in the missing words to complete the following sentences.

(i) Huntington's disease is a disorder of the .... It is caused by a

.... allele. (2 marks)

(ii) Cystic fibrosis is caused by a .... allele. It must be inherited from both parents

The parents might not have the disorder, but they are .... (2 marks)

(Total: .... / 59 marks)

AQA GCSE Science B Workbook Answers

Our Changing Universe (pp 4–7)

1. (a) (i–iii) **In any order:** Reflecting – visible light; Refracting – visible light; Radio – radio waves.
    (b) (i) **Accept suitable answer, e.g.:** Can be used 24 hours a day.
        (ii) **Accept suitable answer, e.g.:** Can only be used at night and if the skies are clear.
    (c) Contains a parabolic glass mirror – Reflecting; Orbits the Earth – Space; Contains two convex (converging) lenses – Refracting; Has a large (non-glass) parabolic dish and a receiver – Radio.
    (d) Light from the object enters; Reflected by a parabolic mirror; Reflected by a flat mirror; Light rays are focused by a convex lens.

2. (a) So they can detect very weak signals. (Radio waves have a much longer wavelength than light waves.)
    (b) (i) Radio telescope.
        (ii) Radio waves are emitted by objects in space; Use a parabolic dish to reflect these waves; Waves reflect to a receiver; The receiver detects and amplifies the signal.

3. (a) They refract light at each end; Using lenses; This causes parallel rays to converge; At a focal point.
    (b) (i) **Accept any two: Refracting**
        Advantages – Portable; Easy to maintain; Cheap (*2 answers needed for 1 mark*)
        Disadvantages – Need visible light; Can only be used at night; Small field of vision. (*2 answers needed for 1 mark*)
        (ii) **Accept any two: Radio**
        Advantages – Can be used 24 hours a day; Can be used to accurately map the Universe; Detect distant/weak signals. (*2 answers needed for 1 mark*)
        Disadvantages – Very expensive to set up; Require a large permanent site. (*2 answers needed for 1 mark*)
    (c) So they are not affected by light pollution and air pollution from built-up areas.

4. (a) Galaxies
    (b) The Doppler effect; The source of light moving away appears to have a longer wavelength than it actually has.
    (c) It tells us that the galaxies are moving away from each other; The Universe is expanding; The further apart the galaxies are, the greater the red-shift.

5. (a) towards; away; expanding
    (b) The further away a galaxy is, the faster it is moving away from us.

6. Row 2 (moving away, moving away)

Our Changing Planet (pp 8–14)

1. (a) Inner core = 2; Outer core = 3; Mantle = 4; Crust = 1
    (b) Rock cycle.
    (c) (i) The jigsaw fit of some continents; Layers of rocks are the same on different continents; The remains of fossils.
        (ii) Continental drift (tectonic theory)

2. (a) The crust and upper mantle
    (b) (i) tectonic plates
        (ii) Convection currents in the mantle

3. (a) intense heat; radioactive decay; mantle
    (b) (i) rock; crust (*both for 1 mark*)
        (ii) down; convection (*both for 1 mark*) (iii) slowly
    (c) 7

4. Slide past each other; Move away from each other; Move towards each other.

5. (a) (i) Destructive (ii) Constructive
    (b) A few centimetres

6. (a) Volcanic eruptions; Earthquakes; Tsunamis
    (b) Destructive
    (c) As two tectonic plates slide past each other; Huge stresses and strains build up in the crust; The eventual release of energy results in an earthquake.
    (d) Because it is situated on a plate boundary.

7. (a) Carbon dioxide
    (b) Nitrogen
    (c) Billions of years
    (d) **Any three from:** Carbon dioxide; Water vapour; Methane; Ammonia; Sulfur dioxide

8. The water vapour in the air condensed.

9. (a) They take in carbon dioxide and give out oxygen.
    (b) It becomes locked up in sedimentary rocks; It is used in the formation of fossil fuels.

10. (a) (i) 200 (*1 mark*)
        (ii) nitrogen; oxygen (*2 marks*)
        (iii) helium (*1 mark*)
    (b) (i) They are unreactive.
        (ii) **Any one from:** Filament lamps; Electric discharge tubes; In balloons

11. (a) photosynthesis; sea; sediment; crust
    (b) (i–ii) **In any order:** Volcanic activity; Burning of fossil fuels.

12. **Any five from:** May lead to substantial climate change; Traditional crops cannot grow; Change in specific physical/abiotic condition, e.g. temperature, day length; A rise in sea level; Loss of areas used for growing food; Loss of areas for housing; Loss of areas for habitats etc.; Extinction of species.

13. (a) (i) Carbon dioxide (ii) Volcanic eruptions
    (b) **Any one from:** Argon; Carbon dioxide; Methane; Water vapour
    (c) **This is a model answer that would score full marks:**
    The levels of three gases have changed over the last 4 billion years. The level of carbon dioxide has decreased and the levels of both oxygen and nitrogen have increased.
    The carbon dioxide level decreased due to several reasons. The evolution of green plants meant that carbon dioxide was taken in during the process of photosynthesis. Carbon dioxide was used up to make carbonates and it was also taken in during the creation of fossil fuels.
    Photosynthesising plants released oxygen into the atmosphere and bacteria released nitrogen from decaying material.

Materials Our Planet Provides (pp 15–23)

1. **(a)** Electron – 3; Electron shell – 2; Nucleus – 1; Atom – 4
   (b) (i) atom
   (ii) shell/energy level
   (iii) nucleus
   (iv) the same

2. **(a)** Sodium – Na
   Oxygen – O
   Iron – Fe
   Sulfur – S
   (b) A substance that consists of only one type of atom.
   (c) 100

3. **(a) (i) In any order:** protons; electrons.
   (both correct for 1 mark)
   (ii) protons/neutrons; neutrons/protons; electrons.
   (All 3 correct for 1 mark)
   (b) Neutron – 0; Proton – +1; Electron – –1.
   (c) They contain the same number of electrons and protons

4. Molecule – A particle that consists of two or more atoms chemically joined
   Element – All of the atoms are the same in these substances
   Atom – The simplest particle that can exist on its own
   Compound – A substance formed when two or more different types of atom are chemically joined together

5. **(a)** 8
   (b) 6
   (c) 2
   (d) 2, 6
   (e) –2

6. They form bonds; Electrons can be gained or lost; Electrons can be shared.

7. **(a)** 4 **(b)** 2 **(c)** NH_3

8. NaOH – 1 molecule, 3 elements *(1 mark)*
   3ZnO – 3 molecules, 2 elements *(1 mark)*
   $2H_2O$ – 2 molecules, 2 elements *(1 mark)*
   $4H_2SO_4$ – 4 molecules, 3 elements *(1 mark)*

9. **(a)** mass; atomic
   (b) protons; protons
   (c) (i) 8
   (ii) C
   (iii) 23
   (iv) 23 – 11 = 12 neutrons
   (v) A proton has the same relative mass as a neutron

10. **(a)** *(1 mark for each correct column)*

	$\begin{array}{c}14\\N\\7\end{array}$	$\begin{array}{c}197\\Au\\79\end{array}$	$\begin{array}{c}235\\U\\92\end{array}$	**(i)** __40__ Ca **(ii)** __20__
Number of protons	**(i)** 7	**(ii)** 79	**(iii)** 92	20
Number of neutrons	**(i)** 7	**(ii)** 118	**(iii)** 143	20

   (b) Atoms of the same element/same atomic number; But with different numbers of neutrons/different mass numbers.

11. **(a)** They have the same atomic number **(b)** 1 **(c)** 1 **(d)** 4 **(e)** isotopes

12. **(a)** 39
    (b) 19
    (c) 19
    (d) It has the same number of protons and electrons.

13. **(a)** Mixture of compounds; Hydrocarbons; Alkanes
    (b) top; bottom
    (c) Short-chain hydrocarbon

14. **(a)** Electrolysis
    (b) Aircraft; Replacement joints
    (c) Water pipes
    (d) gold; compounds; ores
    (e) Any three from; Gold; Silver; Platinum; Copper.

15. **(a)** Redox reaction
    (b) hard
    (c) Carbon; Other metals

16. **(a)** More than one type of substance not chemically joined
    (b) A mixture of hydrocarbon molecules
    (c) True
    (d) The crude oil is evaporated; Then allowed to condense; At a range of different temperatures to form fractions.

17. **(a)** The relative formula mass
    (b) M_r = 56 + 16 = 72 *(1 mark for sum, 1 mark for total)*
    (c) calcium carbonate $\longrightarrow$ calcium oxide + carbon dioxide
    (d) $CaCO_3 \longrightarrow CaO + CO_2$

Using Materials from Our Planet to Make Products (pp 24–28)

1. **(a)** Reactants: magnesium and oxygen *(both needed for 1 mark)*
   Product: magnesium oxide
   (b) (i) Corrosive. **(ii) Accept any suitable answer, e.g.:** Wear safety glasses.
   (c) (i) Type of metal.
   (ii) Accept any correct variable, e.g: Volume of acid added
   (iii) It would increase or decrease the rate of reaction.
   (d) (i) 1 – magnesium; 2 – iron; 3 – copper.
   (ii)

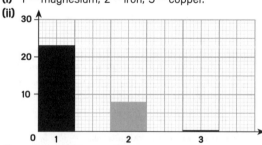

   (1 mark for plotting test tube 1 at 23, 1 mark for plotting test tube 3 at zero)

   (e) (i) Allows ions/particles to move; Weakens forces between molecules
   (ii) Energy is used to break bonds between atoms
   (iii) Accept any two: Large amounts of energy needed; To melt oxide; And split chemical bonds.

2. **(a)** Results show iron will displace copper; Copper does not displace calcium or iron; Calcium displaces iron and copper.
   (b) Calcium is more reactive than iron and copper; Iron is more reactive than copper; Copper is least reactive.

3. **(a) Accept any sensible uses, e.g:** Aluminium used as foil; Gold used as jewellery
   (b) Atoms arranged in rigid framework; Atoms touching many others, allowing energy to pass from one to the next.

4. **(a)** $CH_4 + 2O_2 \longrightarrow CO_2 + 2H_2O$
   (b) $N_2 + 3H_2 \longrightarrow 2NH_3$

5. **(a)** $CuCO_3 \longrightarrow CuO + CO_2$
   (b) $2HCl + CuO \longrightarrow CuCl_2 + H_2O$

6. $Ca(OH)_2 + H_2SO_4 \longrightarrow CaSO_4 + 2H_2O$

7. $2Cu + O_2$ (1 mark) $\longrightarrow 2CuO$ (1 mark)

Life on Our Planet (pp 29–33)

1. **(a)** The total number of individuals of the same species living in a particular habitat.
   (b) A group of animals and plants interacting with one another.
   (c) Food; Water; Space (All correct for 1 mark)
   (d) Accept any suitable answer, e.g: Light
   (e) better adapted (1 mark); survive (1 mark); larger (1 mark).
   (f) It would decrease; Because it would be out-competed by the better adapted organisms.

2. **(a)** A feature that develops to make an organism better suited to its environment.
   (b) It makes an organism better suited to its environment, so more likely to survive.
   (c) Any two from: Thick fur – insulates against the cold; Layer of fat under the skin – insulates against the cold; Small ears – reduces convection and reduces heat loss; White in colour – camouflage; Sharp claws – dig through the snow/ice.
   (d) Reduce water loss; Only stomata on stem/no leaves so no stomata; Prevents/reduces the number of animals that can eat them (The answer 'Reduce surface area for water loss' would be worth 2 marks.)

3. **(a)** 3 billion years ago
   (b) Fossils
   (c) population/species; species; adapted
   (d) Mutation; Change in a gene; Gene codes for protein.

4. **(a) This is a model answer that would score full marks:**
   When the trees were light-coloured, the light-coloured flies were well camouflaged so were more likely to survive and reproduce, compared with the dark-coloured flies. However as the trunks of the trees have gradually become darker, this would have left the light-coloured flies more visible. They would therefore have been more likely to be seen and eaten. Over the years this would have gradually reduced the frequency of the pale wing colour allele.
   The dark-coloured flies would now be more camouflaged against the dark tree trunks, so would be more likely to survive and reproduce, producing dark-coloured offspring.
   (b) Any three from: New/increased competition; Loss of habitat/food; New predators; New diseases; Pollution; Natural disaster.

Biomass and Energy Flow Through the Biosphere (p 34)

1. **(a) (i)** rosebush
   (ii) aphid
   (iii) blackbird
   (b) The Sun
   (c) Accept any two: Keeping warm; Movement; Growth; Respiration; Excretion/egested; Transferred to environment.
   (d) Limit animal movement – less energy needed for muscle contraction (2 marks); Keep barns/buildings warm – less energy needed to maintain body temperature (2 marks)

2. The amount of living matter in an organism.

The Importance of Carbon (pp 35–37)

1. **(a)** remove; growth; returned; die
   (b) In any order: Warmth; Moisture; Oxygen.
   (c) (i) photosynthesis; respiration; decomposition;
   (ii) Carbohydrates, fats and proteins

2. **(a)** ions; soil; roots
   (b) Proteins; DNA/RNA
   (c) chlorophyll
   (d) carbon dioxide + water (1 mark) $\longrightarrow$ glucose + oxygen (1 mark)
   (e) (i) Chlorophyll **(ii)** Inside the chloroplasts

3. Amount of oxygen

4. **(a)** Amount of carbon dioxide; The temperature; Amount of chlorophyll (3 correct = 2 marks, 1–2 correct = 1 mark)
   (b) The lower the rate of photosynthesis will be; Less energy provided for making new chemicals
   (c) Accept two reasons and explanations from: Correct temperature – optimum for enzymes; Carbon dioxide concentration – optimum for growth; Water – needed for cytoplasm/transport.

Control of Body Systems (pp 38–45)

1. Brain; Spinal cord; Paired spinal nerves; Receptors.

2. **(a)** Brain **(b)** Spinal cord **(c)** Neurones

3. **(a)** Neurones are specially adapted cells, which carry electrical signals, e.g. nerve impulses
   (b) (i–ii) In any order: They are long, to make connections around the body (2 marks); Branched endings, which act on a few cells at the same time. (2 marks)

4. The skin

5. synapse; chemical transmitter; receptor; electrical impulse; chemical transmitter.

6. **(a)** sensory; motor; relay.
   (b) True
   (c) A synapse
   (d) A chemical transmitter is released

7. Neurones

8. **(a)** Receptor $\longrightarrow$ Sensory neurone $\longrightarrow$ CNS $\longrightarrow$ Motor neurone $\longrightarrow$ Effector
   (b) Brain
   (c) reflex; response/effector; relay/motor; relay/motor.

9. **(a)** Removing your hand from a hot plate.

(b) Receptor detects stimulus; Travels along sensory neurone; Through a relay; Travels along motor neurone; Effector/muscle/gland responds. *(Max 3 marks)*

(c) Happens very quickly; Protects the body from danger/damage; No thought required.

10. **(a) Any three from:** Eyes – light; Nose – chemicals/smell; Tongue – chemicals/taste; Ears – sound/balance; Skin – pressure/pain.

(b) Coordinates the stimuli and the response.

(c) (i) Test 1 **(ii)** $0.36 + 0.37 + 0.35 + 0.32 = 1.40$
$\frac{1.40}{4}$; $= 0.35$ *(Answer 0.30 gains 1 mark)*.

(d) Test reaction times in a quiet room; Repeat at least 5 times; Provide a regular noise; Compare both sets of results.

(e) Any two from: Distractions reduce concentration levels; Likely to slow down reaction times; More likely to have an accident.

11. **(a)** Keeping our internal environment constant.
(b) Sweating; Breathing out; Urinating.
(c) By eating and drinking

12. **(a)** In glands **(b)** In the bloodstream **(c)** Brain
(d) Oestrogen **(e)** Insulin

13. **(a)** Pancreas does not produce insulin
(b) Insulin; Glucagon

14. **(a)** Heart **(b)** 37°C
(c) Sweat glands stop producing sweat; Shivering occurs; Blood vessels in the skin constrict
(d) Accept any two: Blood vessels in skin dilate; Skin becomes flushed; Heat loss increases; Sweat is produced

15. **(a)** Pancreas
(b) This is a model answer that would score full marks:
Insulin and glucagon help control blood glucose concentration. Blood glucose levels are monitored by the brain.
Insulin acts on glucose and converts it to glycogen, which is released when the levels of glucose in the blood are too high.
Glucagon acts on glycogen and converts it to glucose, which is released when the levels of glucose in the blood are too low.

Chemistry in Action in the Body (p 46)

1. **(a)** acid; alkali
(b) (i) A salt is a metal compound; Made from between a metal and an acid
(ii) Metal + acid; Metal oxide + acid; Metal carbonate + acid

2. Magnesium

3. **(a)** Bases
(b) alkalis

4. **(a)** salt
(b) Sodium nitrate

Human Inheritance and Genetic Disorders (pp 47–52)

1. **(a) A** cell membrane **B** cytoplasm **C** nucleus
D mitochondrion
(b) (i) Nucleus **(ii)** Cell membrane **(iii)** Cytoplasm
(iv) Chromosomes

2. **(a)** Differences between members of the same species.
(b) (i–ii) Genetic – **accept any correct example, e.g:** eye colour.
Environment – **accept any correct example, e.g:** a diet without protein will restrict growth.

3. **(a)** Asexual reproduction; Repair of damaged cells; Growth of new cells *(All three correct for 1 mark)*
(b) chromosome; genetic; parent
(c) 2

4. **(a)** 23 pairs of chromosomes.
(b) It divides repeatedly; By mitosis to form a new individual.
(c) 46
(d) Gametes
(e) Half the number of chromosomes as a normal body cell
(f) A zygote.

5. **(a)** ovaries; eggs and sperm; chromosomes
(b) Gametes

6. **(a)** XX and XY
(b) A
(c) Whether the ovum is fertilised by an X-carrying sperm or a Y-carrying sperm.

7. **(a)** Two
(b) In any order: dominant; recessive.
(c) (i) Homozygous dominant.
(ii) Heterozygous.
(iii) Homozygous recessive.

8. A2; B1; C4; D3

9. **(a) (i)** dominant **(ii)** recessive
(b) Genotypes of the parents; Genotypes of gametes; Genotypes of offspring; Ratio of genotypes.

10. **(a)** Bb
(b) 0%
(c) (i) Parents (BB) (bb)
Gametes (B) (B) (b) (b) *(2 marks)*
Offspring (Bb) (Bb) (Bb) (Bb) *(1 mark)*
(ii) 100

11. **(a)** A disease passed on from parent to child by genes.
(b) nervous system; dominant
(c) recessive; carriers

Materials Used to Construct Our Homes (pp 53–57)

1. **(a)** Sedimentary
(b) Calcium carbonate
(c) Any three from: Building material; Producing slaked lime; Making glass; Making cement, mortar and concrete.
(d) (i) calcium oxide + carbon dioxide *(2 marks)*
(ii) calcium oxide *(1 mark)*
(e) Thermal decomposition

2. **(a) (i)** Calcium carbonate
(ii) Calcium oxide
(iii) Carbon dioxide
(b) calcium oxide; calcium hydroxide; base
(c) To neutralise soil; To prevent crop failure
(d) Slaked lime

3. **(a)** Limestone reacts with acid in rainwater
(b) Concrete; Glass; Mortar; Cement

4. **(a)** Glass – Powdered limestone is mixed with sand; and sodium carbonate; it's then heated.
Cement – Powdered limestone; Roasted; With powdered clay.
Mortar – Cement; Mixed with sand and water.
Concrete – Mortar; Mixed with gravel, sand and water.
   (b) (i) Glass
   (ii) It is transparent.
   (iii) Mortar
   (iv) It sets as a solid.

5. **(a)** The force
   (b) Newton (N)
   (c) Accept two suitable hazards and precautions, e.g:
   Wear safety glasses – chemicals harmful; Mass could fall – Keep feet out of the way

6. **(a) Example:** ethene
   (b) Polymer
   (c) Polymerisation
   (d) Only single bonds; Between carbon atoms
   (e) Poly(ethene) – Bottles
   Polystyrene – Protective packaging
   Poly(propene) – Ropes

Fuels for Cooking, Heating and Transport (pp 58–59)

1. **(a)** Contains carbon and hydrogen only; Carbon atoms joined by single covalent bonds only.
   (b)
   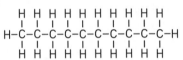
   (2 marks for above, 1 mark for 10 carbon atoms in a chain)
   (c) $x = 8$, $y = 18$
   (d) Carbon atoms are joined to four other atoms; By single bonds

2. **(a)** A substance that releases energy when it is burned.
   (b) Electricity
   (c) Hydrogen

3. **(a)** Carbon dioxide; Water
   (b) sulfur; burn; sulfur dioxide

4. **(a)** carbon dioxide
   (b) (i) True
      (ii) True

Generation and Distribution of Electricity (pp 60–63)

1. **(a)** An energy source that cannot be replaced within a lifetime.
   (b) False
   (c) reactor; fission; thermal; reactor; water; steam; turbines
   (d) Uranium, Plutonium
   (e) fission; collide; split; chain

2. **(a)** Cheap and easy to obtain; Coal-fired power stations are quick to start up.
   (b) It is non-renewable
   (c) They have high decommissioning costs

3. **(a)** Nuclear
   (b) One that will not run out; Can be regenerated
   (c) (i) tides **(ii)** Tidal barrage
   (d) (i) wind **(ii)** Wind turbines

4. **(a)** Hydro-electric power station
   (b) Geothermal

5. **(a)** Increasing the voltage, decreasing the current.
   (b) high; low

6. **(a)** National Grid; voltage; increased; decreased
   (b) stepped down
   (c) Alternating current
   (d) To where it's wanted; In the form that's wanted.
   (e) transformed; wasted

The Cost of Running Appliances in the Home (pp 64–67)

1. A3, B1, C2, D4

2. The energy transformed by a drill is destroyed

3. **(a) (i)** Light **(ii)** Heat
   (b) $\frac{25}{100} \times 100$ *(1 mark)* = 25% *(1 mark)*
   (c) $50 \times \frac{50}{100}$ *(1 mark)* = 25 J/s *(1 mark)*
   (d) 400 − 240 = 160 *(1 mark)*; $\frac{160}{400} \times 100$ *(1 mark)* = 40% *(1 mark)*

4. **(a)** Watts, W
   (b) 3 × 2 × 5 *(1 mark)* = 30p *(1 mark)*
   (c) 7 × 0.5 × 6 *(1 mark)* = 21p *(1 mark)*

5. **(a) (i)** Electricity **(ii)** Heat
   (b) (i) $\frac{20}{25} \times 100$ *(1 mark)* = 80% *(1 mark)*
      (ii) 80J (100 − 20) *(1 mark for 80, 1 mark for J)*
   (c) Any two from: Heat warms up the air around the bulb; Amounts of energy are too small to measure/very difficult to measure; Amounts too small to be used.

Electromagnetic Waves in the Home (pp 68–71)

1. **(a)** Infrared rays – 3; Gamma waves – 1; Radio waves – 4; Ultraviolet rays – 2
   (All correct = 3 marks, 2 or 3 correct = 2 marks, 1 correct = 1 mark)
   (b) All electromagnetic radiation can travel across a vacuum.
   (c) vacuum; particles; energy; frequency
   (d) 300 000 × 1000 *(1 mark)* = 300 000 000 *(1 mark)* m/s *(1 mark)*

2. absorbed; hotter; alternating; frequency

3. A 3, B 1, C 2, D 4

4. Radio waves – High levels of exposure for short periods can increase body temperature, causing tissue damage; Microwaves – Used by mobile phone networks; Infrared waves – Absorbed by the skin and felt as heat; Ultraviolet rays – Used for security coding; X-rays – Give images of broken bones, as they can pass easily through soft tissue; Gamma rays – Used to kill cancerous cells and bacteria on food, but even exposure to low doses can cause cancer.

5. Radio waves; Microwaves; Visible light; Infrared rays.

6. It increases the kinetic energy of the water particles.

7. **(a)** Microwaves are absorbed by water in the cells; The water heats up; Heat denatures proteins; Correct function of proteins in cells, e.g. enzymes, structural.
   (b) (i) Accept five points from: Choose a large sample of people who use mobile phones; Choose a

large sample of people who do not use mobile phones; Calculate percentage of people affected; Control all other variables; Identify at least one control variable; Compare their results with other studies; Compare the results with other group.

(ii) Accept two points from: Mobile phones have not been around for that long; The effect on the next generation is not yet known; There is some idea that genes/DNA may be affected and damage passed on.
(Any sensible answer credited)

The Use (and Misuse) of Drugs (pp 72–77)

1. **(a)–(b)** From natural sources, e.g. plants; From manufactured sources, e.g. synthetic

2. natural; manufactured; useful; animals
    (All correct = 2 marks, 2 correct = 1 mark)

3. **(a)** tested; trialled; toxic; side effects
    (b) Poisonous

4. **(a)** (ii) sleeping pills
    (b) (i) morning sickness
    (c) (iii) limb abnormalities
    (d) (i) comprehensive drug testing
    (e) (iii) leprosy

5. **(a)** Microbes that cause disease.
    (b) Any two from: Pathogens enter the body through wounds that break the skin; Through the respiratory system; By sexual transmission.
    (c) Any two from: White blood cells can engulf pathogens; Release antitoxins; Release antibodies.
    (d) painkillers; destroy; bacteria; viruses.
    (All correct = 2 marks, 2-3 correct = 1 mark)
    (e) This is a model answer that would score full marks: In a population of bacteria, some individuals have natural resistance. This resistance may have been caused by a mutation and is a natural product of variation within a population. Antibiotics kill non-resistant bacteria, but the resistant bacteria survive. The resistant bacteria are then able to reproduce, passing on their resistance. Eventually only the resistance bacteria remain.

6. **(a)** smaller; living; toxins
    (b) Tuberculosis
    (c) Flu
    (d) Platelets
    (e) To destroy toxins
    (f) Produce antibodies; Produce antitoxins; Engulf foreign bodies

7. Aspirin

8. **(a)** painkillers; kill; viruses; reproduce.
    (b) MRSA

9. **(a)** Nicotine
    (b) Any one from: Changes the body's chemistry/reactions; Craved by the body.

10. **(a)** Nicotine patches.
    (b) $\frac{2}{100}$ × 100 *(1 mark)* = 2% *(1 mark)*
    (c) Addictive/changes body chemistry.
    (d) (i–ii) Any two suitable answers from: Age; Gender; Medical health; Fitness.

11. **(a) (i)** At least 5; Can see a pattern. **(ii)** Improves reliability of mean; Allows anomalies to be spotted.

(b) It increases more rapidly at first; The heart rate then reaches a steady rate.
(c) Ethical reasons, e.g. effect may be harmful; Validity, male/female – test same gender.

The Use of Vaccines (pp 78–80)

1. **(a) (i)** False **(ii)** False **(iii)** False **(iv)** True
    (b) Mumps; Measles; Rubella.
    (c) An inactive/dead pathogen
    (d) Vaccine contains a harmless treated pathogen; White blood cells produce antibodies/antitoxins; White blood cells become sensitized; Respond to any future infection more quickly.

2. **(a) Any suitable answers, e.g.:** Wash hands – removes bacteria; Cover cuts – prevents pathogens entering their bodies
    (b) (i) Measure diameter in different directions; calculate a mean **(ii)** $\frac{12 + 13}{2}$ *(1 mark)* = 12.5 *(1 mark)*
    (iii) Either measurement for antibiotic D. **(iv)** Repeat the test for that antibiotic. **(v)** Improves reliability of mean; Reduces effect of variation; Allows anomalies to be identified. **(vi)** Most effective antibiotic will have the largest clear zone around it.

The Use of Ionising Radiation in Medicine (pp 81–83)

1. **(a)** A = Nucleus, B = Electron
    (b) Protons; Neutrons

2. **(a)** isotope; ions; alpha
    (b) (i) Beta
    (ii) Gamma
    (c) A1, B2, C3
    (d) (i) Alpha and beta radiation *(both answers needed for 1 mark)*
    (ii) Because they are charged particles
    (e) It is not deflected by electric fields; It is an electromagnetic wave.
    (f) To track the movement of substances around the patient's body.

3. **(a) In any order:** Sterilisation of food/equipment; Treating cancer
    (b) cancer; greater; inside; outside
    (c) (i) Alpha **(ii)** It cannot penetrate the skin so does not reach cell nucleus.

4. **(a)** Alpha radiation is highly ionising to internal tissue; This means it can damage molecules. It doesn't travel very far inside living cells, but the damaged cells die; This can be beneficial if the cells are cancerous.
    (b) Wear safety badges that measure their exposure – if level gets too high they can stop working in that area; Wear protective clothing, (e.g. lead-lined aprons) and work behind lead glass safety screen – prevents radiation reaching living cells.

Uses of Electroplating (pp 84–86)

1. **(a)** liquid; aqueous *(both correct for 1 mark)*
    (b) The use of electricity to break down a compound; Break a chemical bond.
    (c) Reduction and oxidation reactions take place at the same time.
    (d) ions; reduced; atoms; oxidised.
    (4 correct = 3 marks, 2–3 correct = 2 marks, 1 correct = 1 mark)

2. (a) Chlorine
   (b) $2H^+(aq) + 2e^- \longrightarrow H_2(g)$

3. (a) Looks more attractive; Reduces the risk of allergic reactions.
   (b) Use the item as the negative electrode; Place in a solution containing silver ions; Pass an electric current through the solution; Silver ions deposited on item as silver metal.
   (c) (i) (2 marks for plotting points, 1 mark for line of best fit)

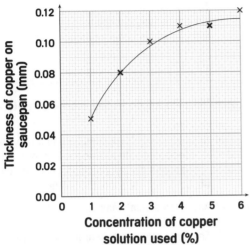

   (ii) As the concentration of the solution increases, the thickness of the copper increases; Increases quickly at first.
   (iii) 4% is the best value for money; Greater concentration will cost more; Higher concentration does not increase thickness as much.

Developing New Products (pp 87–88)

1. (a) The study of very small structures.
   (b) nm
   (c) In any order: Stronger; Stiffer; Lighter.
   (d) Any three suitable answers, e.g.: The car industry; Medical and dental applications; Construction materials.

2. (a) A type of nanostructure that can be designed to have specific properties; And behave in specific ways.
   (b) This is a model answer that would score full marks:
   Smart paint heals its own scratches, which reduces corrosion and keeps a product looking better. When used on products like cars this means that the product holds its value better.
   Super conductors make powerful electromagnets, which have many uses. For example they can be used in MRI scanners to detect illness, brains tumours, etc. They can also be used in Maglev trains, which are less polluting to the environment because no fossil fuels are needed to power them.
   Chromic materials change colour depending on the temperature. They can be used in skin thermometers, which are safe to use. They can also be used in ink used in food packaging, which indicates when items are stored at the correct temperature.

Selective Breeding and Genetic Engineering (pp 89–91)

1. (a) Asexually; Sexually (both correct for 1 mark)
   (b) They are genetically identical to each other; They are genetically identical to the parent plant.

2. (a) Genetically identical to parent/original cell.

   (b) Scrape off small groups of cells; Grow in soil/water/ hydroponics; Add minerals; Add hormones to promote growth.
   (c) They will have the same useful characteristics as the parent; They are genetically identical to the parent plant.

3. (a) True
   (b) False
   (c) True

4. (a) genes; early; characteristics; cloned
   (b) Any three from: To improve crop yield; To improve resistance to pests or herbicides; To extend the shelf-life of fast-ripening crops; To harness the cell chemistry of an organism so it produces a required substance.

5. (a) For; Against; Against; For; For (All correct = 3 marks, 3–4 correct = 2 marks, 2–3 correct = 1 mark)
   (b) First statement
   (c) Third statement
   (d) Accept either the second or fourth statement.

6. This is a model answer that would score full marks:
   Being able to 'design' babies could be advantageous in that it may remove the risk of disease as undesirable alleles will not be passed onto a child. This would reduce the cost to the NHS and reduce the problems to an individual of inheriting a disease that could shorten life expectancy, involve painful medical treatments, etc. A baby could be 'designed' to help treat a disease of an older sibling. It could also help to balance out a family in terms of gender, for example if a couple wants to have a boy and a girl.
   However, being able to choose the gender of a future baby could lead to gender imbalance, as in some cultures boys are more highly valued than girls. There are potential future risks / dangers of this procedure that are not known at the moment. Insurance companies too might use known genetic information to increase premiums or refuse insurance cover.

Environmental Concerns – Making and Using Products (pp 92–95)

1. (a) Less land available for plants and animals; An increase in non-renewable energy sources being used. (Both answers correct for 1 mark)
   (b) With accelerating speed.

2. (a) Nitrogen
   (b) Cause acid rain; Kills trees and plants; Increase carbon dioxide
   (c) Pesticides; Herbicides.
   (d) Any two from: Sewage; Fertiliser; Toxic chemicals.
   (e) For timber; To provide land for agricultural use/building.

3. (a) Carbon dioxide
   (b) Photosynthesis

4. (a) tropical; trees; carbon dioxide; biodiversity; extinct; habitats. (All correct = 4 marks, 5 correct = 3 marks, 3 or 4 correct = 2 marks, 2 correct = 1 mark)
   (b) The range of different organisms; Living in a habitat.

5. (a) improve; generations; global
   (b) Maintain stocks in the long term; At levels where we can use them; But without causing permanent decreases in the resource.
   (c) Accept suitable answer, e.g.: Replace timber; by replanting cut down trees regularly.

6. Any four from: Introduce quotas; Use smaller volume of nets; Closed season; No-fish zones; Increasing mesh size.

7. **(a)** Plant a new sapling for each tree cut down.
    (b) Different species of tree provide: Different food sources; Different habitats; Greater number of food chains; Greater number of different predators.
    (c) Any four from: Greater diversity of habitats; Greater range of food; More insect species; More plant species; Greater number of food chains; Carbon dioxide taken up by plants; Greater levels of photosynthesis; Reduction in carbon dioxide levels in the atmosphere; Carbon dioxide linked to greenhouse effect.

Saving Energy in the Home (pp 96–99)

1. **(a)** Convection
    (b) Conduction

2. **(a)** conduction
    (b) convection
    (c) convection

3. **(a)** Conduction; Metal is a solid; Free electrons surrounded metal atoms; Energy transferred from atom to atom *(Max 3 marks)*
    (b) Convection; Particles spread out and become less dense; Rise and cool down; Cycle starts again.

4. Metals are good conductors because they contain free electrons that are able to move through the metal.

5. convection

6. electromagnetic wave

7. No particles of matter are involved

8. **(a)** Black is a good absorber and emitter; Light, shiny materials are poor absorbers and emitters.
    (b) All objects emit and absorb thermal radiation; The amount of radiation an object gives out or takes in depends on its surface; The amount of radiation an object gives out or takes in depends on its shape and size.

9. faster

10. Put a shiny material (e.g. foil) around it; Insulate it; Use a lid; Decrease surface area.

11. **(a) (i)** Put insulation in
       (ii) Cavity wall insulation
       (iii) Draught excluder/curtains
    (b) (i) 4.30 pm
       (ii) 19°C
       (iii) 7.30 pm; Rate of heat loss reduces.

Controlling Pollution in the Home (pp 100–101)

1. **(a)** Oxygen.
    (b) (i) Combustion produces carbon monoxide; Combines with haemoglobin in red blood cells; Reduces oxygen reaching cells; Can cause death.
       (ii) Less heat released; But still pay same price for gas.

2. **(a) Any three suitable answers:** Dust; Mould; Spores; Pollen; Smoke
    (b) Any three suitable answers: Asthma; Headaches; Tiredness

3. Radon can cause cancer; Radon can be found in houses; Radon is radioactive.

1. **(a)** What type of rock is limestone? Tick the correct option.   (1 mark)

Sedimentary ◯

Metamorphic ◯

Igneous ◯

Natural ◯

(b) What is the main chemical compound in limestone?   (1 mark)

....

(c) Give **three** uses of limestone.   (3 marks)

1. ....

2. ....

3. ....

(d) Limestone reacts in many different ways. Complete the word equations **(a)** and **(b)** in the table below to show two of these reactions.   (3 marks)

(i) limestone	Heat ⟶	.... + ... (quicklime)
(ii)	+ water ⟶	calcium hydroxide (slaked lime)

(e) What type of chemical reaction is used to turn limestone into quicklime?   (1 mark)

....

2. **(a)** Calcium carbonate was heated in a Bunsen burner flame until a reaction took place.

In the reaction that took place, give the name of the…

(i) reactant .... (1 mark)

(ii) solid product .... (1 mark)

(iii) gas produced .... (1 mark)

(b) Ring the correct options in the following sentences: (3 marks)

Limestone can be heated to form **calcium oxide** / **calcium hydroxide**. Water is then added to form **calcium oxide** / **calcium hydroxide**.

Calcium hydroxide is a(n) **acid** / **base**.

(c) Why might a farmer want to add calcium hydroxide to his field? (2 marks)

....

....

(d) What is a common name for calcium hydroxide? (1 mark)

....

3. **(a)** Which of the following is a disadvantage of using limestone as a building material?
Tick the correct option. (1 mark)

Disadvantage	Tick (✓)
Limestone is heavy	
Limestone reacts with acid in rainwater	
It is widely available	
It is easy to shape	

(b) Name four products that can be made from limestone. (4 marks)

1. ....

2. ....

3. ....

4. ....

4. **(a)** Describe how the following substances or materials are made.

Glass: (3 marks)

....

....

....

Cement: (3 marks)

....

....

....

Mortar: (2 marks)

....

....

....

Concrete: (2 marks)

....

....

....

(b) (i) Which limestone product would you use to make windows? (1 mark)

....

(ii) Explain your choice of material for windows. (1 mark)

....

....

(iii) Which limestone product would you use to make bricks stick together? (1 mark)

....

(iv) Explain your choice of material for sticking bricks together. (1 mark)

....

....

5. Kaumal made three different mixtures of concrete and tested them by hanging a load on them and recording the maximum force the concrete could take before it broke.

(a) What is the dependent variable in this investigation? Tick the correct option. (1 mark)

The amount of sand in the concrete ◯

The amount of cement in the concrete ◯

The force ◯

The amount of rock chippings in the concrete ◯

(b) What would be an appropriate unit for the dependant variable? Tick the correct option. (1 mark)

Newton (N) ◯

Kilogram (kg) ◯

Litre (l) ◯

Metre (m) ◯

(c) What potential hazards should Kaumal be aware of in carrying out this practical investigation, and what precautions should she take? (4 marks)

....

....

....

....

....

6. Plastics are large molecules made from oil. Polythene is a plastic.

(a) What is the name of a monomer that is used to make polythene? (1 mark)

....

(b) What type of molecule is formed when monomers are joined together? (1 mark)

....

(c) Name the type of reaction that takes place when monomers are joined together. (1 mark)

....

(d) The molecules formed are often saturated. Explain the meaning of the term **saturated**. (2 marks)

....

....

(e) Draw a line to link each polymer to its specific use. (3 marks)

Poly(ethene)		Protective packaging
Polystyrene		Ropes
Poly(propene)		Bottles

Bottles

Rope

(Total: .... / 51 marks)

1. Decane is an alkane that has the formula $C_{10}H_{22}$.

    (a) How would you describe the chemical structure of an alkane? (2 marks)

    ....

    ....

    ....

    ....

    (b) Draw a diagram to show the arrangements of the atoms in a molecule of decane. (2 marks)

    (c) When decane is heated with a catalyst, the following reaction takes place.

    $$C_{10}H_{22} \longrightarrow C_xH_y + C_2H_4$$

    Use the equation to work out the formula of C_xH_y. (2 marks)

    $x =$ .... $y =$ ....

    (d) 'Alkanes are fairly unreactive.' Use your knowledge of their structure to explain this
    statement. (2 marks)

    ....

    ....

2. (a) What is a fuel? (1 mark)

    ....

    (b) Which of the following is **not** a fuel? Tick the correct option. (1 mark)

    Electricity ⬭   Coal ⬭

    Hydrogen ⬭   Petrol ⬭

    (c) Which fuel produces only water as a waste product? (1 mark)

    ....

3. (a)

Coal

Fossil fuels contain mainly hydrocarbons. What products are produced when a hydrocarbon burns completely? Tick the **two** correct answers. (2 marks)

Carbon ◯

Carbon monoxide ◯

Carbon dioxide ◯

Water ◯

(b) Fill in the spaces in the following sentences by choosing words from the box below. (3 marks)

burn	**sulfur**	**sulfur dioxide**
boils	**oxygen**	**nitrogen**

Many fossil fuels contain .... as an impurity.

As the fuels .... the impurity also reacts with oxygen and produces

....

4. (a) Complete the following sentence:

The greenhouse gas that is the product of burning fossil fuels is .... (1 mark)

(b) Are the following sentences **true** or **false**? (2 marks)

(i) Acid rain concentrations in the atmosphere can be reduced by removing sulfur dioxide from the

waste gases of combustion. ....

(ii) Acid rain can be reduced by removing sulfur from fossil fuels before they are burned.

....

(Total: .... / 19 marks)

1. **(a)** What is the meaning of the term 'non-renewable energy source'? (1 mark)

....

(b) Is the following statement **true** or **false**? (1 mark)

Wood is a non-renewable fuel. ....

(c) Fill in the missing words to complete the following passage. (7 marks)

Nuclear fuel is used to generate electricity. A .... is used to generate heat by

nuclear ....

A heat exchanger transfers .... energy from the .... to

.... The water turns to .... and drives

the ....

(d) Which of the following can be used to make nuclear fuels? Tick the **two** correct options. (2 marks)

Coal ⬭

Uranium ⬭

Propane ⬭

Plutonium ⬭

(e) Circle the correct options in the following sentences. (4 marks)

Nuclear fuels produce energy through a process called **fusion / fission**. Nuclear particles are
released, which **fuse / collide** with the nuclei of other atoms, causing them to **grow / split**. This
causes a **chain / finite** reaction that releases huge amounts of energy.

2. **(a)** Below is a list of statements about using coal as an energy source.

Tick the statements that are advantages of using coal as an energy source. (2 marks)

Statement	Tick (✓)
Cheap and easy to obtain	
Burning produces carbon dioxide	
Coal-fired power stations are quick to start up	
Sulfur in coal contributes to acid rain	

(b) Give **one** disadvantage of using coal as a fuel. (Do not use any answer from part **a**) (1 mark)

....

(c) Which of the following is a valid argument **against** nuclear power stations? Tick the correct option. (1 mark)

Argument	Tick (✓)
They have high fuel costs	
They produce gases that pollute the atmosphere	
They have high decommissioning costs	
They need to be in constant use to work efficiently	

3. Energy resources can be used to generate electricity.

(a) Which one of the following non-renewable energy resources does **not** produce carbon dioxide and sulfur dioxide as waste gases when used in power stations? (1 mark)

Oil ☐

Gas ☐

Nuclear ☐

Coal ☐

(b) What is the meaning of the term 'renewable energy source'? (2 marks)

....

(c) Many renewable energy sources are affected by the Sun or the moon. (2 marks)

(i) Fill in the missing word to complete the following sentence.

The gravitational pull of the moon creates ....

(ii) Name **one** method of generating energy that relies on this.

....

(d) (i) Fill in the missing word to complete the following sentence. (1 mark)

The Sun causes convection currents, which result in ....

(ii) Name **one** method of generating energy that relies on this. (1 mark)

....

4. **(a)** Which type of power station involves the damming of a river flowing in an upland valley?
Tick the correct option. (1 mark)

Power station	Tick (✓)
Nodding duck wave generator	
Tidal barrage	
Geothermal power station	
Hydro-electric power station	

(b) In some volcanic areas hot water and steam rise naturally to the surface of the Earth. The steam can be used to drive turbines. Which type of renewable energy source is this describing?
Tick the correct option. (1 mark)

Energy source	Tick (✓)
Hydro-electric	
Tidal	
Wind	
Geothermal	

5. Electricity is transferred from power stations to our homes along metal wires.

(a) How can energy loss in power lines be reduced? Tick the correct option. (1 mark)

Statement	Tick (✓)
Increasing the current, decreasing the voltage	
Increasing the voltage, decreasing the current	
Increasing current and increasing the voltage	
Decreasing current and decreasing the voltage	

(b) Fill in the missing words to complete the sentence below. (2 marks)

For domestic use, electricity needs to be transmitted at a _____ voltage and a

_____ current.

6. Electricity is generated in power stations.

(a) Choose the correct words from the box below to complete the following sentences. **(4 marks)**

decreased	National Grid	pylon	wattage
voltage	unaltered	distribution grid	increased

The system of pylons that carry the generated electricity from the power station to our

homes is called the .... Transformers are used to change the

.... of the electricity. If a step-up transformer is used, it is

.... If a step-down transformer is used, it is ....

(b) Circle the correct option in the following sentence. **(1 mark)**

Before the electrical energy reaches our homes, its voltage needs to be **stepped up / stepped down**.

(c) What type of electricity do we use in our homes? Tick the correct option. **(1 mark)**

Direct current ⬚

Alternative current ⬚

Alternating current ⬚

AC/DC ⬚

(d) When a device transfers energy, only some of the energy is useful.

What do we mean by **useful**? **(2 marks)**

....

....

(e) Fill in the missing words to complete the following sentences. **(2 marks)**

The remaining energy is .... into a non-useful energy type. It is referred

to as .... energy.

(Total: .... **/41 marks)**

1. These devices transfer electrical energy in different ways to different types of useful energy.

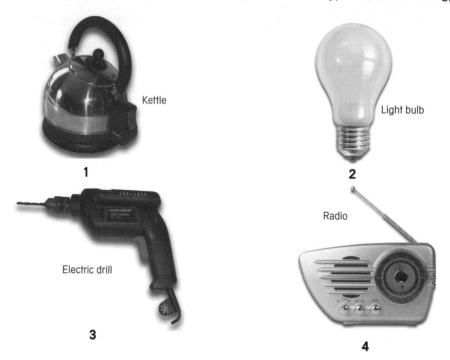

Kettle

1

Light bulb

2

Radio

Electric drill

3

4

Match the types of useful energy **A**, **B**, **C** and **D**, with the devices **1–4** shown above. Enter the appropriate numbers in the boxes provided.   (4 marks)

A   Movement (kinetic) energy   ⬭

B   Heat (thermal) energy   ⬭

C   Light energy   ⬭

D   Sound energy   ⬭

2. Which one of the following statements is **not** true? Tick the correct option.   (1 mark)

Statement	Tick (✓)
The energy transformed by a drill becomes increasingly spread out (dissipated)	
The energy transformed by a drill becomes difficult to use for further energy transformations	
The energy transformed by a drill makes the surroundings a bit warmer	
The energy transformed by a drill is destroyed	

3. **(a)** A standard light bulb is not very efficient. Most of the electrical energy is transformed into a different type of energy.

(i) Name the useful energy type. .... (1 mark)

(ii) Name the 'wasted' energy type. .... (1 mark)

(b) A hairdryer has an input of 100J, and 25J/s is transformed usefully to another form of energy.

What is the device's efficiency? Show your working. (2 marks)

....

....

Answer ....

(c) A fluorescent light works at 50% efficiency.

If it has 50J of electrical energy input, what will be its light energy output?
Show your working. (2 marks)

....

....

Answer ....

(d) A laptop can convert 400J of electrical energy to 240J of useful light and sound.

What percentage of the energy is converted into non-useful (or wasted) energy?
Show your working. (3 marks)

....

....

Answer ....

4. **(a)** Name the unit used to measure electrical power. (1 mark)

....

(b) A 3kW electric heater is switched on for two hours.

How much does it cost to use if a unit of electricity costs 5p? Show your working. (2 marks)

Electric heater

....

....

Answer ....

(c) A 7kW electric shower is used for 30 minutes. What is the cost of using this if a unit of
electrical energy costs 6p? Show your working. (2 marks)

....

....

Answer ....

5. This is an advertisement for a new type of light bulb that claims to be more energy-efficient than other types.

(a) (i) Name the type of energy that goes into a light bulb to make it work. (1 mark)

(ii) Name the type of energy that is 'wasted'. (1 mark)

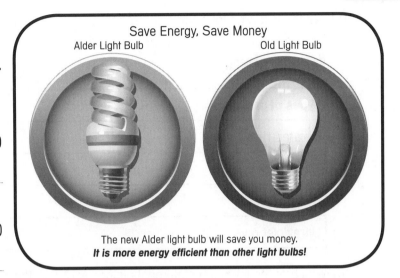

Save Energy, Save Money

Alder Light Bulb   Old Light Bulb

The new Alder light bulb will save you money.
It is more energy efficient than other light bulbs!

(b) The table below shows energy transfer for different light bulbs.

Type of light bulb	Energy in (J)	Useful energy out (J)	Wasted energy out (J)	Efficiency of light bulb (%)
Low energy light bulb	25	20	5	
Standard light bulb	100	20		20

(i) Calculate the efficiency of the low energy light bulb. Show your working. (2 marks)

Efficiency .... %

(ii) How much energy is wasted by the standard light bulb? Show your working. (2 marks)

Answer ....

(c) Some people say 'The energy is lost'. This is not correct. How would you explain what actually happens? (2 marks)

(Total: .... /27 marks)

1. (a) Put the following **four** types of electromagnetic radiation in order of increasing length of wavelength. Write the numbers in the boxes provided (where 1 has the shortest wavelength and 4 has the longest wavelength).

    (3 marks)

Types of radiation	Order
Infrared rays	
Gamma waves	
Radio waves	
Ultraviolet rays	

(b) Which of the following statements is correct? Tick the correct option. (1 mark)

No electromagnetic radiation can be seen by the human eye. ◯

All electromagnetic radiation is visible to the human eye. ◯

All electromagnetic radiation can travel across a vacuum. ◯

Only some parts of the electromagnetic spectrum can travel across a vacuum. ◯

(c) Choose the correct words from the box below to complete the following sentences. (4 marks)

X-rays	frequency	vacuum	particles
	intensity	energy	gamma

All electromagnetic waves can travel across a .... because they do not need

...., unlike sound waves.

All electromagnetic waves transfer ....

Each type of electromagnetic radiation has a different wavelength and

....

(d) A wave has a wavelength of 1000m and a frequency of 300 000Hz. Calculate its speed. Show your working. (3 marks)

....

....

....

Answer ....

2. Circle the correct options in the following sentences. (4 marks)

When a wave is **reflected / refracted / absorbed** by a substance it makes the substance **move / hotter / colder / dissolve**.

It may create an **alternating / direct / modified** current of the same **wavelength / amplitude / frequency** as the radiation.

3. The statements describe four uses of electromagnetic radiation. Match the types of electromagnetic wave **A–D** below with the statements **1–4**. Enter the appropriate numbers in the boxes provided. (4 marks)

1 Transmitting radio signals

2 Treating certain cancers

3 Transmitting satellite television channels

4 Security coding valuables

A Microwaves ⬜

B Radio waves ⬜

C X-rays ⬜

D Ultraviolet rays ⬜

4. Draw a line to match each part of the electromagnetic spectrum in **List A** to the correct description in **List B**. One has been done for you. (5 marks)

List A	List B
Radio waves	Give images of broken bones, as they pass easily through soft tissue.
Microwaves	Used to kill cancerous cells and bacteria on food, but even exposure to low doses can cause cancer.
Infrared rays	Absorbed by the skin and felt as heat.
Ultraviolet rays	Used by mobile phone networks.
X-rays	Used for security coding.
Gamma rays	High levels of exposure for short periods can increase body temperature, causing tissue damage.

5. Which parts of the electromagnetic spectrum are normally used for communications? (4 marks)

1. ....

2. ....

3. ....

4. ....

6. Microwaves are absorbed by the water particles in food. What happens to the energy from the microwaves? Tick the correct option. (1 mark)

Option	Tick (✓)
It is emitted as light or sound energy	
It increases the potential energy of the water particles	
It increases the kinetic energy of the water particles	
It decreases the kinetic energy of the water particles	

7. Some people are concerned about the possible risks to our health when using a mobile phone. Their concerns include the following:

• Some studies have linked mobile phone use with brain tumours.

• The long-term effects of using mobile phones are not known.

(a) Microwaves are the type of electromagnetic waves used by mobile phones. Describe and explain how microwaves can damage living cells. (4 marks)

....

....

....

....

....

....

....

....

(b) Some studies suggest that the use of mobile phones is linked to the growth of brain tumours.

(i) Suggest how reliable data could be collected to test this theory. It would not be ethical to expose people to microwaves as part of the study. (5 marks)

(ii) 'The long term effects of using mobile phones are not known.' Explain the meaning of this statement. (2 marks)

(Total: _____ /40 marks)

1. Describe two ways in which new drugs can be found. (2 marks)

(a) ....

(b) ....

2. Complete the following sentences about drugs using the words in the box below. (2 marks)

animals	manufactured	natural	useful

Some drugs are obtained from .... substances. Others are synthetic, which

means they are .... Drugs can have .... qualities,

but they can also be harmful. Before drugs are trialled using volunteers, they are tested

in the laboratory on ....

3. **(a)** Fill in the missing words to complete the sentences below. (4 marks)

New medical drugs must be .... and .... to find out

whether they are .... They must then be checked for ....

(b) What does toxic mean? (1 mark)

....

4. The statements below describe how a drug called thalidomide was developed.

Underline the correct answer – **(i)**, **(ii)** or **(iii)** – to complete sentences **(a)** to **(e)**. (5 marks)

(a) Thalidomide was approved as…

(i) a cure for cancer   **(ii)** sleeping pills   **(iii)** a pain killer

(b) Thalidomide was then found to relieve…

(i) morning sickness   **(ii)** travel sickness   **(iii)** altitude sickness

(c) Many babies born to mothers who took thalidomide had…

(i) brain damage   **(ii)** poor eyesight   **(iii)** limb abnormalities

(d) The thalidomide example highlights the importance of…

(i) comprehensive drug testing   **(ii)** genetic testing   **(iii)** developing new drugs

(e) Thalidomide is now used to treat…

(i) morning sickness   **(ii)** insomnia   **(iii)** leprosy

5. **(a)** What is a pathogen? (1 mark)

....

(b) Describe **two** ways in which pathogens can enter the body. (2 marks)

1. ....

2. ....

(c) Describe two ways in which white blood cells defend the body against disease. (2 marks)

1. ....

2. ....

(d) Complete the sentences below using the following words. (2 marks)

viruses	**painkillers**	**destroy**	**bacteria**

The symptoms of a disease can often be relieved by using .... However,

these drugs do not .... pathogens. Antibiotics can be used to kill harmful

.... but they cannot kill ....

(e) *In this question you will be assessed on using good English, organising information clearly and using specialist terms where appropriate.*

Overuse of antibiotics can lead to the development of bacteria that are resistant to antibiotics.

Describe the sequence of events that could lead to the development of resistant bacteria. (6 marks)

....

....

....

....

....

....

....

....

....

6. **(a)** Circle the correct words in the following sentences. (3 marks)

Viruses are **smaller / bigger** than bacteria. They can reproduce quickly inside **living / dead** cells.

They produce **toxins / antitoxins**, and cause illnesses.

(b) Which of the following is a disease caused by a bacterium? Tick the correct option. (1 mark)

Ringworm ⬜

Polio ⬜

Measles ⬜

Tuberculosis ⬜

(c) Which of the following is a disease caused by a virus? Tick the correct option. (1 mark)

Tetanus ⬜

Flu ⬜

Athlete's foot ⬜

Ringworm ⬜

(d) Which part of our blood forms scabs over wounds? Tick the correct option. (1 mark)

Red blood cells ⬜

Plasma ⬜

Platelets ⬜

White blood cells ⬜

(e) What is the function of antitoxins? (1 mark)

....

....

(f) What are the functions of white blood cells? (3 marks)

....

....

....

....

....

....

7. Which of the following is an example of a painkiller? Tick the correct option. (1 mark)

Alcohol ⬭

Aspirin ⬭

Insulin ⬭

Vitamins ⬭

8. Drugs can be used to treat us when we are ill.

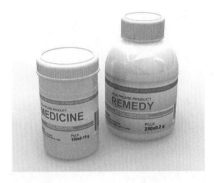

(a) Fill in the missing words to complete the sentences below. (4 marks)

The painful symptoms of a disease can often be alleviated using drugs called

.... However, these drugs do not .... pathogens.

Antibiotics can be used to kill harmful bacteria, but they cannot kill ...., which

live and .... inside cells.

(b) Give an example of a strain of bacteria that has developed resistance to antibiotics. (1 mark)

....

9. **(a)** Name the chemical in cigarette smoke that is addictive. (1 mark)

....

(b) Explain the meaning of 'addictive.' (1 mark)

....

10. Scientists investigated the effect of four different ways to give up smoking: nicotine patches, acupuncture, hypnosis and 'cold turkey' (just stopping smoking straight away). The scientists included 100 people in each trial group. Each method was used for two months.

The results of the investigation are shown below.

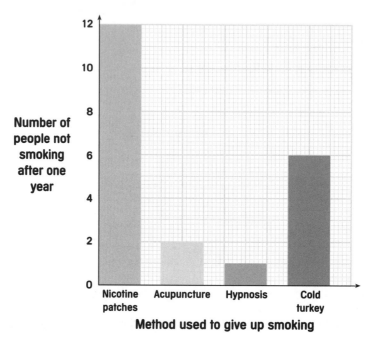

(a) Which method was most successful? (1 mark)

_____

(b) What percentage of the trial group gave up smoking by using acupuncture? Show your working. (2 marks)

Answer _____ %

(c) What effect does nicotine have on the body? (1 mark)

_____

(d) Suggest **two** variables that the scientists would try to control when they chose the people to take part in the investigation. (2 marks)

(i) _____

(ii) _____

11. Coffee, tea and some soft drinks contain caffeine. Caffeine is a drug that affects our bodies. Simon and Cheryll were asked to investigate the effect of different concentrations of caffeine on heart beat. They decided to test the effect of caffeine on the heart beat of water fleas. Water fleas are small animals that live in fresh water. It is possible to see their heart through their body.

(a) (i) How many different concentrations of caffeine should they use? Give a reason for your answer. (2 marks)

_____

_____

(ii) They decided to repeat the test for each concentration three times. Explain why. (2 marks)

_____

_____

(b) The graph below shows some of their results.

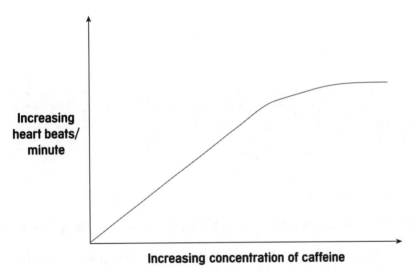

Increasing heart beats/ minute

Increasing concentration of caffeine

Describe what the graph shows about the relationship between increasing concentration of caffeine and heart beats/minute. (2 marks)

_____

_____

(c) Simon and Cheryll made sure that the water fleas were not harmed during this test. They were not allowed to try this investigation on themselves. Suggest **two** reasons why. (2 marks)

_____

_____

(Total: _____ / 61 marks)

1. **(a)** Write **true** or **false** alongside each of the following statements about vaccines. (4 marks)

    (i) Red blood cells produce antibodies. ....

    (ii) Antibiotics destroy antibodies. ....

    (iii) A live pathogen is injected into your body. ....

    (iv) Vaccines provide an acquired immunity. ....

(b) Name the three illnesses that the MMR vaccine protects us against. (3 marks)

    (i) .... **(ii)** .... **(iii)** ....

(c) What is the part of a vaccine that is specific to the disease you are trying to prevent? (1 mark)

....

(d) Describe how a vaccine can protect us against a disease caused by a microorganism. (4 marks)

....

....

....

....

....

....

2. Russell and Simon investigated the effect of five different antibiotics on the growth of bacteria.

(a) Describe and explain **two** safety precautions they should take before they start this investigation. (4 marks)

....

....

....

....

....

....

(b) Russell and Simon were given three petri dishes containing nutrient agar and samples of bacteria. Each petri dish contained the same concentration of bacteria. Each plate was evenly covered by bacteria on the surface. The bacteria growing on the surface made the surface look pale grey all over. A small disc of filter paper soaked in antibiotic A was placed on the agar plate. Four other discs soaked in the four other antibiotics B, C, D and E being tested were also placed on the agar plate.

The diagram below shows the appearance of one of the petri dishes.

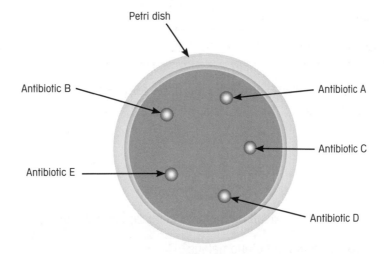

The petri dishes were labelled and incubated at 25°C for three days.

The diagram below shows the appearance of one petri dish after incubation. If the antibiotic kills the bacteria, a clear zone appears around the disc.

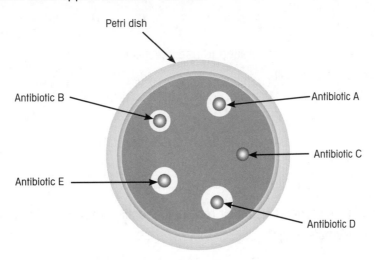

(i) Describe how you would work out the diameter of the clear zone containing antibiotic D. (2 marks)

....

....

(ii) Table B below shows the results for the other two petri dishes.

Table B

		Diameter of clear zone (mm)				
		Antibiotic A	Antibiotic B	Antibiotic C	Antibiotic D	Antibiotic E
Diameter of clear zone (mm)	Dish 2	22	12	2	18	8
	Dish 3	24	13	3	33	9
Mean diameter (mm)		23		2.5		8.5

Calculate the mean diameter of the clear zone for antibiotic B using the results in table B. Show your working. (2 marks)

Mean diameter of clear zone for antibiotic B = ...

(iii) Which results appears to be anomalous? (1 mark)

....

(iv) What action could the students take to check the reliability of the results for antibiotic A? (1 mark)

....

(v) Students are always encouraged to use several results to calculate a mean. What effect does this have on their results? (3 marks)

....

....

....

(vi) How could the students use their results to identify the most effective antibiotic against the bacteria they have used? (1 mark)

....

....

(Total: .... / 26 marks)

The Use of Ionising Radiation in Medicine U3

1. The diagram below represents the structure of an atom.

(a) Fill in the labels **A** and **B** on the drawing of the atom. (2 marks)

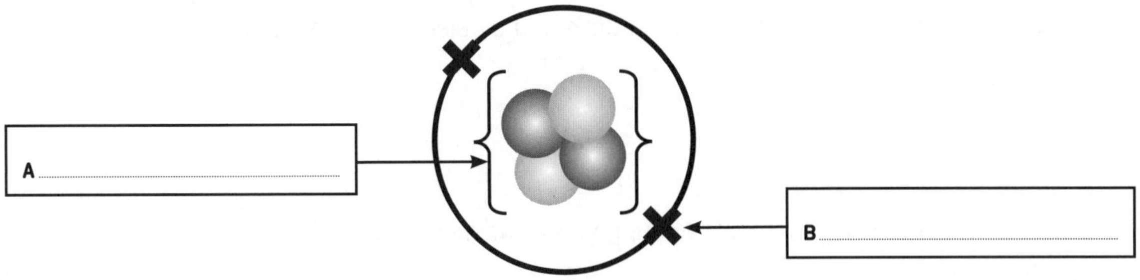

A ....

B ....

(b) Which particles does the nucleus of an atom contain? (2 marks)

....

2. Isotopes are different forms of the same element. Some isotopes are radioactive.

(a) Circle the correct options in the following sentences. (3 marks)

A form of an element with the same number of protons but a different number of neutrons is called an **ion / isotope**.

When a radioactive particle collides with an atom, it can remove electrons and therefore create **ions / isotopes**.

A radioactive particle containing 2 protons and 2 neutrons is an **alpha / beta** particle.

(b) (i) What type of radiation consists of high energy electrons emitted from the nucleus? (1 mark)

....

(ii) What type of radiation is not a particle but an electromagnetic wave? (1 mark)

....

(c) Match statements **A**, **B**, and **C** with the absorbers **1–3** below. Enter the correct number in the box provided. (3 marks)

1 Thick concrete

2 Paper

3 A few cm of aluminium

A The necessary material for absorbing gamma radiation. ◯

B All that is needed to absorb alpha radiation. ◯

C All that is needed to absorb beta radiation. ◯

(d) (i) Which types of radiation are deflected by electric fields and magnetic fields? (1 mark)

....

(ii) Why can these types of radiation be deflected by electric and magnetic fields? (1 mark)

....

....

(e) Which of the following statements about gamma radiation are true?
Tick the correct options. (2 marks)

It is not deflected by electric fields. ☐

It is deflected by magnetic fields. ☐

It is an electromagnetic wave. ☐

It travels faster than visible light in a vacuum. ☐

(f) It is sometimes necessary to inject hospital patients with radioactive isotopes.
What is the purpose of injecting the isotope into the patient? Tick the correct option. (1 mark)

To track the patient's movements around the hospital. ☐

To make the patient sleepy in preparation for an operation. ☐

To track the movement of substances around the patient's body. ☐

To find out the patient's blood group. ☐

3. Radioactive isotopes have many uses but they can also be harmful.

(a) Give **two** common uses of gamma radiation. (2 marks)

....

....

(b) Fill in the missing words to complete the sentences below. (4 marks)

Radiation damage to cells can cause .... The larger the dose of

radiation, the .... the risk.

The damaging effect of radiation depends on whether the source is .... or

.... the body.

(c) (i) Which type of radiation is not a problem to humans if the source is outside the body?   (1 mark)

....

(ii) Explain your answer.   (1 mark)

....

....

4. **(a)** Although exposure to radiation carries a risk of cancer, alpha radiation is often used to treat mouth cancer. Explain why.   (4 marks)

....

....

....

....

....

....

....

....

(b) Workers who may be exposed to radioactivity are trained to work safely to reduce their risk of exposure.

Describe **two** ways in which people who work with radioactivity can reduce the risks.

Explain how each of these ways will reduce their risks.   (4 marks)

....

....

....

....

....

....

(Total: .... / 33 marks)

1. Electrolysis can be used to extract and purify metals and to plate them.

   (a) Circle the correct options in the following sentence. (1 mark)

   To undergo electrolysis, ionic compounds must be **solid / liquid** or **gas / aqueous**.

   (b) What does the word **electrolysis** mean? (2 marks)

   ....

   ....

   ....

   ....

   (c) What does the term 'redox reaction' mean? (1 mark)

   ....

   ....

   (d) Circle the correct options in the following sentences. (3 marks)

   In the electrolysis of molten zinc chloride, zinc **ions / atoms** are **oxidised / reduced** to zinc

   ions / atoms at the negative electrode. At the positive electrode, negatively charged ions

   are **oxidised / reduced**.

2. Sodium chloride is an ionic compound that can undergo electrolysis to produce useful chemicals.

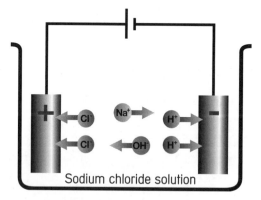

Sodium chloride solution

   (a) Name the substance produced at the positive electrode. (1 mark)

   ....

   Higher Tier

   (b) Hydrogen gas is released at the negative electrode. Write a symbol equation for the reaction
   that takes place. (1 mark)

   ....

3. Some metals are plated with silver or gold.

(a) Give **two** reasons for plating nickel metal with silver or gold. (2 marks)

....

....

(b) Describe how a ring could be plated with silver. You may wish to draw a diagram to help explain your answer. (4 marks)

....

....

....

(c) Copper metal is a good conductor of heat. The thicker the copper plating the better it is at conducting heat, but the more expensive it is to produce. A company called Sally Engineering produce copper-plated saucepans. The rising costs of plating materials with copper was causing Sally Engineering to lose money.

Two scientists who worked for the company were asked to find a more cost-effective way of plating the saucepans with copper.

The scientists investigated the process of copper plating saucepans and obtained the following results.

Concentration of copper salt solution (%)	Thickness of copper on saucepan (mm)
1	0.05
2	0.08
3	0.10
4	0.11
5	0.115
6	0.12

(c) (i) Use these results to complete the graph below. Draw a line of best fit. Some points have already been plotted.

(3 marks)

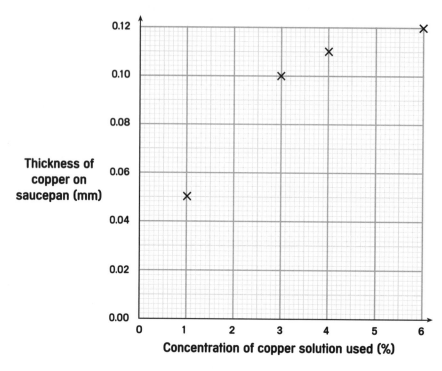

(ii) Use the graph to describe the pattern in the results. (2 marks)

....

....

....

....

(iii) Write a report to give to the company to suggest the concentration of copper solution that would give the best 'value for money'. Give reasons for your choice. (3 marks)

....

....

....

....

....

....

(Total: .... / 23 marks)

1. **(a)** What is nanoscience? (1 mark)

....

....

(b) Circle the correct options in the following sentence. (1 mark)

The unit for a nanometre is **m / c / nm / mn**.

(c) Describe **three** advantages nanocomposites have over plastics. (3 marks)

1. ....

2. ....

3. ....

(d) Describe **three** ways in which nanotechnology is used in industry. (3 marks)

1. ....

2. ....

3. ....

2. **(a)** Describe the meaning of the term 'smart material'. (2 marks)

....

....

....

....

(b) *In the following question you will be marked on your ability to use good English, organise information clearly and use specialist terms where appropriate.*

Describe the uses and advantages of smart materials.

(6 marks)

....

....

....

....

....

....

....

....

....

....

(Total: .... / 16 marks)

Selective Breeding and Genetic Engineering U3

1. **(a)** Plants can reproduce naturally in two ways. What are these two ways?
   Tick the correct boxes. (1 mark)

   Sexually ☐

   By genetic modification ☐

   Asexually ☐

   By embryo transplantation ☐

   (b) Which of the following statements about the offspring produced by asexual reproduction are true?
   Tick the correct options. (2 marks)

   They show genetic variation. ☐

   They are genetically identical to each other. ☐

   They are genetically identical to the parent plant. ☐

   They are infertile. ☐

2. **(a)** Explain the meaning of the word a **clone**. (1 mark)

   ....

   ....

   (b) Many new plants can be produced artificially by cloning. Describe how new carrot plants can be
   produced by cloning. (4 marks)

   ....

   ....

   ....

   ....

   ....

   ....

   (c) Explain the advantages of producing new carrot plants by cloning. (2 marks)

   ....

   ....

   ....

   ....

3. Are the following statements **true** or **false**? Write your answers in the spaces provided. (3 marks)

   (a) A tissue culture technique produces offspring that are genetically identical to the parent plant. ....

   (b) Embryo transplantation produces offspring that are identical to the host female. ....

   (c) In adult cell cloning, the DNA from a donor animal is inserted into an egg cell that has had its nucleus removed. ....

4. (a) Circle the correct options in the following sentences. (4 marks)

   Genetic modification is a process in which **genes/hormones** from one organism are transferred into another.

   The genes are often transferred at a **early / late** stage of development, so that the organism will develop with the desired **behaviour / characteristics**.

   More organisms with the same characteristics can be produced if the genetically modified organism is **cloned / reproduced** with another member of the same species.

   (b) Give **three** reasons why genetic modification is useful in the production of food crops. (3 marks)

   1. ....

   2. ....

   3. ....

5. Embryos produced by IVF can now be screened for a range of genetic diseases. Only healthy embryos are implanted into the uterus and develop into babies.

   (a) The statements in the table below either support this screening or are against it. Put a tick in the **for** or **against** column next to each statement. (3 marks)

Statement	For	Against
The money currently spent on treating these diseases could be spent on more urgent medical care		
Disease is a natural way of controlling population		
It would devalue the life of people who already have this disease		
It would prevent a lot of pain and suffering		
It would free up hospital beds and resources for other patients		

(b) Select **one** statement from part **(a)** that is concerned with economics (money). (1 mark)

_____

(c) Select **one** statement from part **(a)** that is concerned with ethics. (1 mark)

_____

(d) Select **one** argument from part **(a)** that is concerned with society (overall effect on population). (1 mark)

_____

6. _In this question you will be marked on your ability to use good English, organise information clearly and use specialist terms where appropriate._

'Designer babies' is a term used to describe the idea that parents could 'choose' the characteristics they want their baby to have. Some people agree with this idea and others don't.

What are the advantages and disadvantages of allowing people to select the characteristics of their baby? (6 marks)

_____

_____

_____

_____

_____

_____

_____

_____

_____

(Total: _____ / 32 marks)

1. **(a)** Which two of the following effects have been created by an increase in the human population? Tick the **two** correct options. (1 mark)

   Less land available for plants and animals ⬜

   Less waste being produced ⬜

   A reduction in pollution ⬜

   An increase in non-renewable energy sources being used ⬜

   (b) The human population is increasing exponentially. What does **exponentially** mean? (1 mark)

   ....

2. **(a)** Which of the following gases is normally present in large amounts in unpolluted air? Underline the correct answer. (1 mark)

   Oxygen

   Carbon dioxide

   Methane

   Nitrogen

   (b) Which of the following is a way that waste gases from a coal-fired power station can cause air pollution? Tick the correct options. (3 marks)

   Cause acid rain ⬜

   Decrease global warming ⬜

   Kill trees and plants ⬜

   Increase carbon dioxide ⬜

   (c) Name **two** chemicals used in farming that can pollute the soil. (2 marks)

   1. ....

   2. ....

   (d) Name **two** chemicals that can affect the water in our rivers and lakes. (2 marks)

   1. ....

   2. ....

(e) Give **two** reasons for large-scale deforestation. (2 marks)

....

....

3. **(a)** What is the name of the gas released when trees are burned? Tick the correct option. (1 mark)

Carbon dioxide ☐

Oxygen ☐

Methane ☐

Hydrogen ☐

(b) Which biological process decreases the amount of carbon dioxide in the atmosphere? Tick the correct option. (1 mark)

Respiration ☐

Exhalation ☐

Photosynthesis ☐

Deforestation ☐

4. Human activities change habitats and ecosystems in many ways.

(a) Circle the correct option(s) in the sentences below. (4 marks)

When deforestation occurs in **tropical / arctic / desert** regions, it has a devastating impact on the environment.

The loss of **trees / animals / insects** means less photosynthesis takes place, so less

oxygen / nitrogen / carbon dioxide is removed from the atmosphere.

It also leads to a reduction in **variation / biodiversity / mutation**, because some tree species may

become **devolved / damaged / extinct** and **habitats / land / farms** are being destroyed.

(b) Explain the meaning of the word 'biodiversity'. (2 marks)

....

....

....

....

5. **(a)** Put a (ring) round the correct options in the following paragraph. (3 marks)

Sustainable development ensures that development can take place to help **improve / reduce / compromise** or sustain quality of life, without compromising the needs of future **space travel / generations / mutations.** It is an important consideration at local, regional and **sea / carbon dioxide / global** levels.

(b) Explain the meaning of the term 'sustainable development'. (3 marks)

....

....

....

(c) Briefly explain how **one** resource can be managed sustainably. (2 marks)

....

....

6. Fish as a food source provides us with protein, vitamins, minerals and fats.

Describe how we can maintain ocean fish stocks. (4 marks)

....

....

....

....

....

7. Forests and woodlands are being destroyed by human activities. The different species of trees found in woodland provide many useful substances.

(a) Describe **one** way in which forests and woodlands can be managed sustainably. (1 mark)

(b) What is the key principle behind sustainable forest management? (2 marks)

(c) Describe and explain the effect of maintaining large areas of different species of trees on biodiversity. (4 marks)

(Total: .... / 39 marks)

1. Circle the correct options in the following sentences. (2 marks)

   (a) Conduction / Convection is the transfer of heat energy through the movement of a liquid or gas.

   (b) Conduction / Convection is the transfer of heat energy through a solid without the substance itself moving.

2. Fill in the missing words to complete the following sentences. (3 marks)

   (a) The process of heat energy being transferred along a metal bar is called _____.

   (b) The process of heat energy being transferred through a moving liquid is called

   _____.

   (c) The process of heat energy being transferred through a moving gas is called

   _____.

3. Some soup is cooked in a metal container on a camp fire.

Metal cook pot

Fire

   (a) How is the thermal energy transferred through the metal of the container to the soup? (3 marks)

   _____

   _____

   _____

   _____

   _____

   (b) How is the thermal energy transferred through the soup? (4 marks)

   _____

   _____

   _____

   _____

   _____

4. Which of these statements about metals are **not** correct? (1 mark)

Statement	Tick (✓)
Metals are good conductors because they contain free electrons that are able to move through the metal	
Metals are good conductors because they do not have free electrons that are able to move through the metal	
As the metal becomes hotter, its particles gain more kinetic energy	
As the metal becomes hotter, the free electrons move faster	

5. Circle the correct option in the following sentence. (1 mark)

A radiator heats the air in a room mainly by **radiation / convection / conduction**.

6. Fill in the missing words to complete the sentence below. (1 mark)

Infrared radiation is the transfer of heat energy using a type of ....

....

7. How is the transfer of heat by infrared radiation different from the transfer of heat by conduction and convection? (1 mark)

....

8. **(a)** Which of the following statements about the emission and absorption of infrared radiation are true?

Tick the **two** correct options. (2 marks)

Statement	Tick (✓)
Black is a good absorber and emitter	
Black is a poor absorber and emitter	
Black is a poor absorber and good emitter	
Black is a good absorber and poor emitter	
Light, shiny materials are good absorbers and emitters	
Light, shiny materials are poor absorbers and emitters	
Light, shiny materials are poor absorbers and good emitters	
Light, shiny materials are good absorbers and poor emitters	

(b) Which of the following statements about thermal radiation are true?

Tick the **three** correct options. (3 marks)

All objects emit and absorb thermal radiation.

The cooler the object, the more energy it radiates.

The amount of radiation an object gives out or takes in depends on its surface.

The amount of radiation an object gives out or takes in depends on its shape and size.

9. Circle the correct option in the following sentence. (1 mark)

An object will emit or absorb energy **faster / slower** if there's a big difference in temperature between it and its surroundings.

10. Philipa and Viv want to keep their drinks hot when they are outside.

Suggest how they can slow down the rate at which heat is transferred to keep their drinks hotter for longer. (4 marks)

11. Heat is transferred from inside our home to the outside.

(a) Describe how heat transfer can be reduced from the: (3 marks)

(i) Roof ....

(ii) Walls ....

(iii) Doors ....

(b) The graph shows the temperature inside a room of a house between 4pm and 8pm.

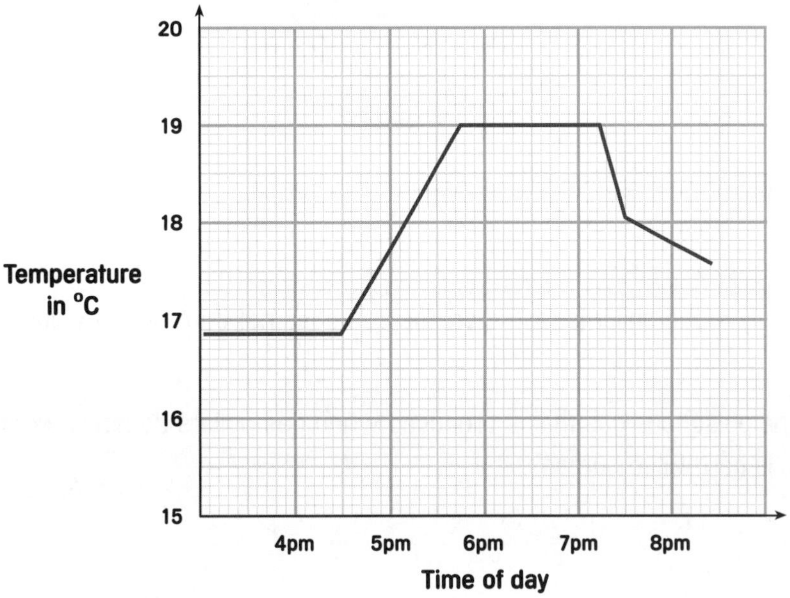

(i) What time did the heating switch on? (1 mark)

....

(ii) What was the highest temperature in the room? (1 mark)

....

(iii) The heating switched off at 7pm. Jon closed the curtains to reduce heat loss through the windows.

What time do you think he closed the curtains? Explain why you think he closed the curtains at this time. (2 marks)

....

....

....

(Total: .... / 33 marks)

1. Gas heating boilers burn natural gas as a fuel to release heat.

Boiler

(a) Name the gas from the air that is used when fuels burn. (1 mark)

....

(b) If the air supply to the boiler is restricted, the gas does not burn properly and harmful products are released.

(i) Name the poisonous gas that is produced when gas boilers do not work properly. Explain how this gas is harmful to humans. (4 marks)

....

....

....

....

....

....

....

(ii) Gas boilers that do burn gas properly cost more to heat our homes. Explain how. (2 marks)

....

....

....

....

2. **(a)** Name **three** common pollutants that might be found in our homes. (3 marks)

1. ....

2. ....

3. ....

(b) Pollutants found in our homes can make us ill. (3 marks)

Describe **three** symptoms caused by high levels of indoor pollution.

1. ....

2. ....

3. ....

3. Radon gas is sometimes detected in houses. Some of the statements below are true and some are false. Put ticks in the boxes next to the true statements. (3 marks)

Statement	True (✓)
Radon can cause cancer	
Radon can be found in houses	
Radon is a solid	
Radon is radioactive	
Soil and rocks contain radium and uranium	

(Total: .... **/ 16 marks)**

Notes

Data Sheet

Power = Potential difference × Current

$$\text{Power} = \frac{\text{Energy transferred}}{\text{Time}}$$

Total cost = Number of kilowatt-hours × Cost per kilowatt-hour

$$\text{Efficiency} = \frac{\text{Useful energy out}}{\text{Total energy in}}$$

$$\text{Efficiency} = \frac{\text{Useful power out}}{\text{Total power in}}$$

Velocity = Frequency × Wavelength

The Periodic Table

Key

relative atomic mass
atomic symbol
name
atomic (proton) number

	1	2

													3	4	5	6	7	0

1
H
hydrogen
1

Group 1
- 7 **Li** lithium 3
- 23 **Na** sodium 11
- 39 **K** potassium 19
- 85 **Rb** rubidium 37
- 133 **Cs** caesium 55
- [223] **Fr** francium 87

Group 2
- 9 **Be** beryllium 4
- 24 **Mg** magnesium 12
- 40 **Ca** calcium 20
- 88 **Sr** strontium 38
- 137 **Ba** barium 56
- [226] **Ra** radium 88

Transition metals

Period 4	Period 5	Period 6	Period 7
45 **Sc** scandium 21	89 **Y** yttrium 39	139 **La*** lanthanum 57	[227] **Ac*** actinium 89
48 **Ti** titanium 22	91 **Zr** zirconium 40	178 **Hf** hafnium 72	[261] **Rf** rutherfordium 104
51 **V** vanadium 23	93 **Nb** niobium 41	181 **Ta** tantalum 73	[262] **Db** dubnium 105
52 **Cr** chromium 24	96 **Mo** molybdenum 42	184 **W** tungsten 74	[266] **Sg** seaborgium 106
55 **Mn** manganese 25	[98] **Tc** technetium 43	186 **Re** rhenium 75	[264] **Bh** bohrium 107
56 **Fe** iron 26	101 **Ru** ruthenium 44	190 **Os** osmium 76	[277] **Hs** hassium 108
59 **Co** cobalt 27	103 **Rh** rhodium 45	192 **Ir** iridium 77	[268] **Mt** meitnerium 109
59 **Ni** nickel 28	106 **Pd** palladium 46	195 **Pt** platinum 78	[271] **Ds** darmstadtium 110
63.5 **Cu** copper 29	108 **Ag** silver 47	197 **Au** gold 79	[272] **Rg** roentgenium 111
65 **Zn** zinc 30	112 **Cd** cadmium 48	201 **Hg** mercury 80	

Group 3
- 11 **B** boron 5
- 27 **Al** aluminium 13
- 70 **Ga** gallium 31
- 115 **In** indium 49
- 204 **Tl** thallium 81

Group 4
- 12 **C** carbon 6
- 28 **Si** silicon 14
- 73 **Ge** germanium 32
- 119 **Sn** tin 50
- 207 **Pb** lead 82

Group 5
- 14 **N** nitrogen 7
- 31 **P** phosphorus 15
- 75 **As** arsenic 33
- 122 **Sb** antimony 51
- 209 **Bi** bismuth 83

Group 6
- 16 **O** oxygen 8
- 32 **S** sulfur 16
- 79 **Se** selenium 34
- 128 **Te** tellurium 52
- [209] **Po** polonium 84

Group 7
- 19 **F** fluorine 9
- 35.5 **Cl** chlorine 17
- 80 **Br** bromine 35
- 127 **I** iodine 53
- [210] **At** astatine 85

Group 0
- 4 **He** helium 2
- 20 **Ne** neon 10
- 40 **Ar** argon 18
- 84 **Kr** krypton 36
- 131 **Xe** xenon 54
- [222] **Rn** radon 86

Elements with atomic numbers 112–116 have been reported but not fully authenticated

*The lanthanoids (atomic numbers 58–71) and the actinoids (atomic numbers 90–103) have been omitted.
The relative atomic masses of copper and chlorine have not been rounded to the nearest whole number.